AF576036

Recent Advances in Cobalt and Related Catalysts

Recent Advances in Cobalt and Related Catalysts: From Catalyst Preparation to Catalytic Performance

Editor

John Vakros

MDPI • Basel • Beijing • Wuhan • Barcelona • Belgrade • Manchester • Tokyo • Cluj • Tianjin

Editor
John Vakros
Chemistry
University of Patras
Patras
Greece

Editorial Office
MDPI
St. Alban-Anlage 66
4052 Basel, Switzerland

This is a reprint of articles from the Special Issue published online in the open access journal *Catalysts* (ISSN 2073-4344) (available at: www.mdpi.com/journal/catalysts/special_issues/Co_catal).

For citation purposes, cite each article independently as indicated on the article page online and as indicated below:

LastName, A.A.; LastName, B.B.; LastName, C.C. Article Title. *Journal Name* **Year**, *Volume Number*, Page Range.

ISBN 978-3-0365-1422-2 (Hbk)
ISBN 978-3-0365-1421-5 (PDF)

© 2021 by the authors. Articles in this book are Open Access and distributed under the Creative Commons Attribution (CC BY) license, which allows users to download, copy and build upon published articles, as long as the author and publisher are properly credited, which ensures maximum dissemination and a wider impact of our publications.
The book as a whole is distributed by MDPI under the terms and conditions of the Creative Commons license CC BY-NC-ND.

Contents

About the Editor vii

John Vakros
Recent Advances in Cobalt and Related Catalysts: From Catalyst Preparation to Catalytic Performance
Reprinted from: *Catalysts* **2021**, *11*, 420, doi:10.3390/catal11040420 1

Mi Wang, Jian Ma, Haoqi Yang, Guolong Lu, Shuchen Yang and Zhiyong Chang
Nitrogen and Cobalt Co-Coped Carbon Materials Derived from Biomass Chitin as High-Performance Electrocatalyst for Aluminum-Air Batteries
Reprinted from: *Catalysts* **2019**, *9*, 954, doi:10.3390/catal9110954 5

Elijah T. Adesuji, Esther Guardado-Villegas, Keyla M. Fuentes, Margarita Sánchez-Domínguez and Marcelo Videa
Pt-Co_3O_4 Superstructures by One-Pot Reduction/Precipitation in Bicontinuous Microemulsion for Electrocatalytic Oxygen Evolution Reaction
Reprinted from: *Catalysts* **2020**, *10*, 1311, doi:10.3390/catal10111311 15

Michela Martinelli, Sai Charan Karuturi, Richard Garcia, Caleb D. Watson, Wilson D. Shafer, Donald C. Cronauer, A. Jeremy Kropf, Christopher L. Marshall and Gary Jacobs
Substitution of Co with Ni in Co/Al_2O_3 Catalysts for Fischer–Tropsch Synthesis
Reprinted from: *Catalysts* **2020**, *10*, 334, doi:10.3390/catal10030334 37

Eduard Karakhanov, Anton Maximov, Anna Zolotukhina, Vladimir Vinokurov, Evgenii Ivanov and Aleksandr Glotov
Manganese and Cobalt Doped Hierarchical Mesoporous Halloysite-Based Catalysts for Selective Oxidation of *p*-Xylene to Terephthalic Acid
Reprinted from: *Catalysts* **2019**, *10*, 7, doi:10.3390/catal10010007 65

John Vakros
The Influence of Preparation Method on the Physicochemical Characteristics and Catalytic Activity of Co/TiO_2 Catalysts
Reprinted from: *Catalysts* **2020**, *10*, 88, doi:10.3390/catal10010088 81

Andrea Bellmann, Christine Rautenberg, Ursula Bentrup and Angelika Brückner
Determining the Location of Co^{2+} in Zeolites by UV-Vis Diffuse Reflection Spectroscopy: A Critical View
Reprinted from: *Catalysts* **2020**, *10*, 123, doi:10.3390/catal10010123 97

About the Editor

John Vakros

Dr. John Vakros serves as a researcher in the Department of Chemical Engineering, University of Patras, Greece. He holds a Diploma (1992) in Chemistry from University of Patras, Greece, and a Ph.D. (1997) from University of Patras. His teaching activities (Hellenic Open University, University of Patras and University of Peloponnese) cover general chemistry, physical chemistry, chemical instrumental analysis, and catalysis at undergraduate and postgraduate levels. His research activities focus on heterogeneous catalytic processes, catalyst preparation, acid–base behaviour of solids, preparation and evaluation of biochars and, over the last five years, the circular economy, with emphasis on catalytic measurements for environmental applications, determination of interactions of reactants with catalytic surface groups and tuning the physicochemical properties of supported catalysts.

Editorial

Recent Advances in Cobalt and Related Catalysts: From Catalyst Preparation to Catalytic Performance

John Vakros

Department of Chemistry, University of Patras, 26504 Patras, Greece; vakros@chemistry.upatras.gr

In our days cobalt, Co, and related catalysts are very attractive because they exhibit a lot of advantages like low cost and high activity in a significant variety of different applications. Cobalt-catalysts are among the most active catalysts for Fischer-Tropsch synthesis and they promote the catalytic activity of the hydrodesulfurization catalysts. Also, they have other significant applications in environmental protection like oxidation of volatile organic compounds, VOC, persulfate activator, ammonia synthesis, electrocatalysis, and many more. Cobalt-catalysts are active, stable, and exhibit significant oxidation-reduction activity, as Co can be found either as Co(II) or Co(III). Also, many molecules can interact with cobalt-supported phase by co-ordination due to partially filled d-orbital. Co-catalysts can be supported in almost all the inorganic supports like alumina, titania, zeolites etc. The cobalt oxide phase can be stabilized onto the surface of the support due to variable interactions between support and cobalt phase. These interactions are crucial for catalytic activity and can be regulated by proper selection of the preparation parameters like the type of support, the Co loading, impregnation method, and thermal conditions.

The energy problem seems to be one of the most serious among the top 10 problems of the humanity. It is more complex since the public discussion about energy is dominated with the climate change. One possible solution is the development of more green approaches for the replacement of fossil fuels. One promising candidate is the development of batteries as clean energy storage and conversion devices. A platinum-based electrocatalyst supported on carbon form is commonly applied in these devices, with the disadvantage of the high cost. It has been reported that the replacement of the expensive Pt with Co and doping of carbon support with N and S heteroatoms can lead to electrocatalysts with excellent performance toward oxygen reduction reaction, oxygen evolution reaction (OER), and hydrogen evolution reaction.

Citation: Vakros, J. Recent Advances in Cobalt and Related Catalysts: From Catalyst Preparation to Catalytic Performance. *Catalysts* **2021**, *11*, 420. https://doi.org/10.3390/catal11040420

Received: 16 March 2021
Accepted: 24 March 2021
Published: 25 March 2021

Publisher's Note: MDPI stays neutral with regard to jurisdictional claims in published maps and institutional affiliations.

Copyright: © 2021 by the author. Licensee MDPI, Basel, Switzerland. This article is an open access article distributed under the terms and conditions of the Creative Commons Attribution (CC BY) license (https://creativecommons.org/licenses/by/4.0/).

Indeed, the work of Wang et al. [1] clearly showed that a Co, N co-doped carbon material (CoNC) can be competitive of the Pt/C electrocatalyst. The carbon support was prepared from chitin and the Co and N co-doped biocarbon was synthesized by a one-step pyrolysis method, which is a simple and low-cost method. The CoNC exhibits the positive onset potential of 0.86 V vs. RHE and high-limiting current density of 5.94 mA cm^{-2}. The stability of CoNC is higher in alkaline electrolyte compared to the commercial Pt/C catalyst. The combination of CoNC as cathode with Al-air battery, provides the power density of 32.24 mW cm^{-2}. These results are promising for the mass production of low-cost ORR catalyst and the utilization of biomass product.

On the other hand the simultaneous presence of cobalt oxides species with Pt metal species can be favorable for the anodic OER reaction. Generally, the OER is characterized by a high onset potential, and it exhibits slow kinetics due to four-electron transfer and oxygen—oxygen bond formation. These factors were the trigger for many researchers to study the substitution of Pt with other oxides. Among them Co_3O_4 is a promising alternative, since it has affordable cost, high abundant in nature, and possess enhanced catalytic performance toward OER in alkaline media.

Indeed, Adesuji et al. [2] have prepared an effective Pt-Co_3O_4 electrocatalyst. Their results showed that the Pt-Co_3O_4-HSs has better performance than Pt-HS. The performance of the new electrocatalyst strongly depends on the preparation method. The authors use bicontinuous microemulsions (BCME) for the synthesis of hierarchical superstructures (HSs) of Pt-Co_3O_4 by reduction/precipitation. They use both hydrophilic and lipophilic precursors and prepared PtAq-CoAq, PtAq-CoOi, PtOi-CoAq, and PtOi-CoOi samples (where Aq and Oi stand for the precursor present in aqueous or oily phase, respectively). The lowest onset overpotentials were achieved with PtOi-CoOi (1.46 V vs. RHE). Also, the PtAq-CoAq exhibits the lowest overpotential at a current density of 10 mA cm^{-2}, equal to 381 mV, significantly lower than the one for Pt-HS (627 mV). The higher Tafel slope was 157 mV dec^{-1} for PtOi-CoAq. The improvement in OER activity is due to the synergistic effect of electronegative Pt on Co_3O_4, which allows for the formation of Co(IV) intermediate species.

One other interesting application, where the Co has central role, is the Fischer-Tropsch synthesis (FTS). This is a catalytic process where syngas is converted to high-quality fuels. The catalytic elements for this process are Co, Fe, Ni, and Ru. Only Fe and Co are used commercially, since Ru is impractical due to very high cost and low abundance and Ni exhibits higher selectivity for the short-chained hydrocarbons, something that it is not desirable. Cobalt has high activity and selectivity for linear long-chained hydrocarbons, low deactivation rate, and finally, low activity for water-gas shift reaction. The activity of a cobalt catalyst depends on the dispersion and surface availability of Co(0) species, with an optimal particle size of 6–10 nm for Co particles. The partial substitution of Co species with Ni can form alloys with different electronic structure with better performance in the FTS.

In the study of Martinelli et al. [3] the effect of cobalt substitution with nickel was systematically investigated for the FTS. Catalysts with different Ni/Co ratios were prepared, characterized with various physicochemical techniques, and tested in a continuously stirred tank reactor for more than 200 h. The authors have found that substitution of Co with Ni did not significantly modify the morphological properties of the mixed catalyst. There is intimate contact between nickel and cobalt, strongly suggesting the formation of a Co-Ni solid oxide solution in each case. Moreover, TPR-XANES indicated that nickel addition improves the cobalt reducibility by systematically shifting the reduction profiles to lower temperatures. The FTS tests revealed that the bimetallic system has a lower initial activity. However, carbon monoxide conversion continuously increased with time on-stream until a steady-state value (34–37% depending on Ni/Co ratio). This value is very close to the undoped Co/Al_2O_3 catalyst. Due to the Ni hydrogenation activity the bimetallic catalysts exhibited slightly higher initial selectivity for light hydrocarbons. Taking in mind the difference in price between cobalt and nickel, the substitution of Co with Ni in a ratio Ni/Co equal to 25/75 can lead to a better performance catalyst compared to the activity/price value.

Different bimetallic catalysts with Co can also be applied in several applications. In the work of Karakhanov et al., [4] a bimetallic MnCo catalyst, supported on the mesoporous hierarchical MCM-41/halloysite nanotube composite, was synthesized and tested for the selective oxidation of *p*-xylene to terephthalic acid (TPA). The MnCo catalyst was prepared by wetness impregnation method using the acetic salts of Mn(II) and Co(II) ions with a nominal molar ratio of Mn/Co equal to 1/10 and real value 1/8.6. Both metals were present as oxides and they interact with the support surface groups. The catalyst was tested in the liquid-phase oxidation of *p*-xylene, a multi stage radical oxidation process. Its kinetics is strongly affected by the reaction conditions. As a rule, *p*-xylene conversion to *p*-toluic acid proceeds fast, whereas oxidation to TPA is a slow highly activated process. The tests were performed under the best condition of AMOCO process. Both homogeneous and heterogeneous catalysts give a quantitative conversion and similar product distribution with TPA yield of about 95% within 3 h but with lower Mn and Co content for the solid catalyst. Comparing the TOF number heterogeneous catalyst has 3.8 times higher TOF value proving the high efficacy and superiority of the heterogeneous catalyst. The reaction pathway is passed through the intermediate alcohol formation. The intermediated radicals are not deactivated with Co(II) species due to their high interactions with the support surface groups, while the intermediate products can be further oxidized by O_2 of Br radicals catalyzed by the Co(II) species.

The interactions of Co(II) species with the support surface groups are very important for the physicochemical characteristics and the activity of the supported Co catalysts. For example the interactions of the Co(II) ions with the surface –OH groups of the Al_2O_3 can change the local structure of the deposited Co species. This is also valid for the case of TiO_2. Indeed the interfacial deposition of Co(II) species on the titania surface has been investigated and it was found that the mode of Co(II) ions deposition depends on the surface coverage. In the case of titania the main deposition mechanism is the formation of mononuclear/oligonuclear inner sphere complexes. No interfacial precipitation occurs in contrast with the Co(II)-Al_2O_3 system. This mode of deposition leads to low content of Co catalysts. In a recent publish paper [5] it was shown that the proper regulation of the impregnation conditions can increase the Co content. Specifically, the application of high initial basicity in the impregnation solution and high equilibration time can increase the Co content. The physicochemical characteristics of the prepared catalyst were superior of the corresponding catalyst prepared with conventional impregnation method. The Co oxide phase is more reducible, exhibit higher degree of interaction with the surface –OH groups, different ratio of Co(III)/Co(II) species and higher surface coverage. The above characteristics result in a more active catalysts for the degradation of sulfamethaxazole using sodium persulfate as oxidant.

The importance of the different local structure of the Co species on the surface of different supports was pointed out in the study of Bellmann et al. [6]. In this study, different supports were used for the preparation of Co-supported catalysts. Specifically, ZSM-5, Na-ZSM-5, Al-SBA-15, and Al_2O_3-SiO_2 were used. UV-Vis spectroscopy and FTIR spectroscopy of pyridine and CO adsorption were applied to determine the nature of Co species in the above materials. It was found that the all Co-zeolite samples contain two types of Co(II) species located at exchange positions as well as in oxide-like clusters independent of the Co content, while in Co/Al-SBA-15 and Co/Al_2O_3-SiO_2 only Co(II) species in oxide-like clusters occur. Finally, the UV-Vis spectra exhibit that not exclusively isolated Co(II) species were present and the characteristic triplet band is not only related to γ-, β-, and α-type Co(II) sites in the zeolite but also to those dispersed on the surface of different oxide supports. These observations clearly denote that for proper characterization of the formed Co species, the use of complementary methods is required.

It is clear that the cobalt catalysts can be used in broad applications. Among their advantages is the low-moderate price and the high abundance of Co. The deposition of Co(II) species in different supports leads to the preparation of highly active and dispersed catalysts. The local environment of the deposited Co species can be regulated through the preparation conditions and stabilized by the interactions with the support surface sites. The applications of cobalt catalysts is not limited to those presented in this book but it is a very active field of research, with promising results. On the other hand, further research is highly needed, especially to obtain better insight into the different processes, in order to develop more active catalysts tailored to the reaction conditions.

Finally, I would like to express my sincerest thanks to all authors for their valuable contributions, as well as the editorial team of *Catalysts* for their kind support and fast responses. Without them, this special issue would not have been possible.

Funding: This research received no external funding.

Acknowledgments: Caroline Zhan is gratefully acknowledged for her valuable help in all the stages of the special issue.

Conflicts of Interest: The author declares no conflict of interest.

References

1. Wang, M.; Ma, J.; Yang, H.; Lu, G.; Yang, S.; Chang, Z. Nitrogen and Cobalt Co-Coped Carbon Materials Derived from Biomass Chitin as High-Performance Electrocatalyst for Aluminum-Air Batteries. *Catalysts* **2019**, *9*, 954. [CrossRef]
2. Adesuji, E.T.; Guardado-Villegas, E.; Fuentes, K.M.; Sánchez-Domínguez, M.; Videa, M. Pt-Co3O4 Superstructures by One-Pot Reduction/Precipitation in Bicontinuous Microemulsion for Electrocatalytic Oxygen Evolution Reaction. *Catalysts* **2020**, *10*, 1311. [CrossRef]
3. Martinelli, M.; Karuturi, S.C.; Garcia, R.; Watson, C.D.; Shafer, W.D.; Cronauer, D.C.; Kropf, A.J.; Marshall, C.L.; Jacobs, G. Substitution of Co with Ni in Co/Al2O3 Catalysts for Fischer–Tropsch Synthesis. *Catalysts* **2020**, *10*, 334. [CrossRef]
4. Karakhanov, E.; Maximov, A.; Zolotukhina, A.; Vinokurov, V.; Ivanov, E.; Glotov, A. Manganese and Cobalt Doped Hierarchical Mesoporous Halloysite-Based Catalysts for Selective Oxidation of p-Xylene to Terephthalic Acid. *Catalysts* **2020**, *10*, 7. [CrossRef]
5. Vakros, J. The Influence of Preparation Method on the Physicochemical Characteristics and Catalytic Activity of Co/TiO2 Catalysts. *Catalysts* **2020**, *10*, 88. [CrossRef]
6. Bellmann, A.; Rautenberg, C.; Bentrup, U.; Brückner, A. Determining the Location of Co2+ in Zeolites by UV-Vis Diffuse Reflection Spectroscopy: A Critical View. *Catalysts* **2020**, *10*, 123. [CrossRef]

Article

Nitrogen and Cobalt Co-Coped Carbon Materials Derived from Biomass Chitin as High-Performance Electrocatalyst for Aluminum-Air Batteries

Mi Wang [1], Jian Ma [2], Haoqi Yang [2], Guolong Lu [2,*], Shuchen Yang [1,*] and Zhiyong Chang [2]

[1] Engineering College, Changchun Normal University, Changchun, Jilin Province 130022, China; wangmi3214@126.com

[2] Key Laboratory of Bionic Engineering (Ministry of Education), College of Biological and Agricultural Engineering, Jilin University, Changchun, Jilin Province 130022, China; majian@126.com (J.M.); yhq1214@126.com (H.Y.); chanzhiyong@126.com (Z.C.)

* Correspondence: Guolonglu@jlu.edu.cn (G.L.); Ysc2017@mail.cncnc.edu.cn (S.Y.)

Received: 21 October 2019; Accepted: 10 November 2019; Published: 14 November 2019

Abstract: Development of convenient, economic electrocatalysts for oxygen reduction reaction (ORR) in alkaline medium is of great significance to practical applications of aluminum-air batteries. Herein, a biomass chitin-derived carbon material with high ORR activities has been prepared and applied as electrocatalysts in Al-air batteries. The obtained cobalt, nitrogen co-doped carbon material (CoNC) exhibits the positive onset potential 0.86 V vs. RHE (reversible hydrogen electrode) and high-limiting current density 5.94 mA cm^{-2}. Additionally, the durability of the CoNC material in alkaline electrolyte shows better stability when compared to the commercial Pt/C catalyst. Furthermore, the Al-air battery using CoNC as an air cathode catalyst provides the power density of 32.24 mW cm^{-2} and remains the constant discharge voltage of 1.17 V at 20 mA cm^{-2}. This work not only provides a facile method to synthesize low-cost and efficient ORR electrocatalysts for Al-air batteries, but also paves a new way to explore and utilize high-valued biomass materials.

Keywords: electrocatalyst; oxygen reduction reaction; Al-air battery; biomass; nitrogen-doped carbon

---

1. Introduction

Environmental deterioration and the depletion of finite resources stimulate research to seek and develop clean energy storage and conversion devices for the future. Metal (Al, Li, Zn, etc.)-air batteries [1–3] are considered the most promising candidates to replace the current fossil fuels due to their low carbon emissions, high theoretical specific capacity and energy density, etc. Among the batteries listed above, the Al-air battery is most abundant in the Earth's crust and possesses a high energy density (8.10 Wh g^{-1}) [4], which gives the Al-air battery considerable potential to be widely used; however, its poor electrochemical performance constrains its practical application.

Oxygen reduction reaction (ORR) is a key reaction affecting the discharged performance of energy storage devices such as fuel cells and metal-air batteries [5–7]. In order to enhance the discharged performance of these devices, a platinum-based electrocatalyst is commonly applied to overcome the sluggish ORR kinetics and accelerate the reaction process [8–11]. Despite the fact that platinum (Pt)-based catalysts are considered the most efficient ORR electrocatalyst, their expensive price and poor durability has limited their practical application [12–14]. Thus, great efforts have been devoted to exploring low-cost and high-performance carbon-based electrocatalysts. Among various carbon materials, graphene [15], carbon nanotubes (CNTs) [16] and activated carbon [17] are most frequently used in the preparation of carbon-based electrocatalysts due to their excellent physical and chemical properties. Although these carbon materials have shown remarkable catalytic activity, their cost still limits their application. Therefore, it is important to seek another substitute for synthesizing a low-cost, carbon-based ORR electrocatalyst.

Biomass materials are the ideal carbon precursor for synthesizing ORR catalysts due to their renewability and large-scale and low-cost production [18–25]. Recently, several studies reported biomass-derived ORR catalysts with efficient catalytic activities. Li and co-workers [26] reported a low-cost method for transforming woody biomass into a single-atom, dispersed ORR catalyst. Wu et al. [27] synthesized kelp-derived, multilayer-graphene-encapsulated cobalt nanoparticles, which could possess efficient and stable electrocatalytic properties. Wei et al. [28] reported a reed-stalk-based Si-Fe/N/C ORR catalyst inspired by the Si-promoting graphitization findings. Deng et al. [29] described an activated carbon sheet derived from chitin, which was an efficient non-metal bifunctional electrocatalyst for both ORR and oxygen evolution reaction (OER). Li et al. [30] designed three-dimensional, honeycomb-like, nitrogen-doped carbon nanosheet/graphene nanonetwork films, which were prepared by dissolution and coagulation of chitin and graphene oxide in NaOH/urea aqueous solution using a repeated freezing–thawing process. Deng et al. [31] reported a chitin-derived porous carbon loaded with Co, N, and S heteroatoms, which showed excellent performance towards ORR, OER, and hydrogen evolution reaction (HER).

As numerous works reported, synthesizing non-precious ORR catalysts with high catalytic activity often requires further treatments such as activation [32] or hydrothermal [33] and chemical vaporous deposition [34]. However, intricate production steps are unpractical in industry and cannot achieve mass production. Herein, a one-step pyrolysis method was developed for preparation of a Co and N co-doped biocarbon catalyst derived from chitin. The obtained biomass, carbon-based material exhibits high catalytic activity and superior durability, which is comparable with the state-of-the-art noble-metal catalyst (Pt/C). The outstanding ORR performance can be attributed to the synergetic effect of the N-content functional group, the Co nanoparticles, and the carbon framework. Furthermore, an Al-air battery was assembled with a Co and N co-doped biocarbon electrocatalyst as an air cathode, which achieved high power density (32.24 mW cm^{-2}) and excellent energy storage performance, indicating the Co and N co-doped biocarbon catalyst could easily be utilized as an efficient electrocatalyst in the Al-air battery.

2. Results and Discussion

2.1. Morphological and Structural Characterization

The morphology of as-prepared catalysts was first characterized by SEM and TEM. As shown in Figure 1a and Figure S1, a large number of pores are randomly distributed in the carbon skeleton of NC and CoNC. Furthermore, numerous particles are distributed in and on the skeleton of CoNC. In the TEM image shown in Figure 1b, several nanoparticles ranging from 15 nm to 40 nm can be observed in CoNC. From HR-TEM in Figure 1c, these nanoparticles are encapsulated in the CoNC skeleton and possess lattice spacing of 0.212 nm corresponding to Co particles. Raman spectra were applied to assess the defects of as-prepared materials. As shown in Figure 1d, Raman spectra of the NC and CoNC both have two obvious peaks (denoted as D band and G band) located in 1350 cm^{-1} and 1590 cm^{-1}, which correspond to local defects of carbon materials and graphite degree graphitic in-plane vibration of the carbon materials, respectively. The intensity ratio of D band and G band (I_D/I_G) indicates the defect degree of carbon materials. As is shown in Figure 1g, the I_D/I_G of the CoNC and NC is about 1.06 and 0.95, respectively, which suggests that the CoNC materials contain more defects in the carbon skeleton than those of NC. Subsequently, the difference in morphology was further discussed via a nitrogen adsorption-desorption measurement. Based on nitrogen adsorption-desorption isotherms (Figure 1e), it can be noted that the CoNC possessed higher specific surface areas (~165 m^2 g^{-1}) than that of NC (~106 m^2 g^{-1}). Likewise, the CoNC also showed the lowest average pore size (2.99 nm), lower than that of NC (5.47 nm) (Figure 1f). The possible reason for the above phenomenon might be ascribed to the sharp shrinking of the carbon matrix and catalytic effect induced by transition metal during the pyrolysis process [35,36].

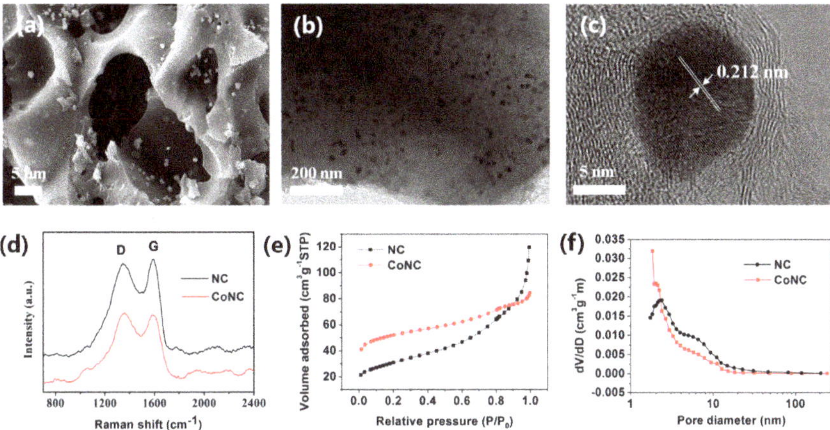

Figure 1. The SEM and TEM images of cobalt nitrogen co-doped carbon material (CoNC) (**a**–**c**). Raman spectra (**d**), N_2 adsorption/desorption isotherms (**e**) and the corresponding pore size (**f**).

The chemical states of as-prepared catalysts were studied by X-ray photoelectron spectroscopy (XPS). As shown in Figure 2a, the spectrum of CoNC contains four significant peaks with locations at 285 eV, 533 eV, 400 eV, and 782 eV corresponding to the C, O, N, Co elements, respectively. These results corresponded to the observation of the SEM and TEM images, suggesting that there are Co particles in CoNC. In a high-resolution C1s spectra of CoNC (Figure 2b), the detected peaks can be deconvoluted into 284.6 eV, 286.4 eV, 289.1 eV, and 285.4 eV, and each of them indicates the existence of C-C bond, C=O bond, O-C=O, and C-N bond in CoNC, respectively. Among these element peaks, the existence of a C-N bond implies the successful doping of N elements into the carbon skeleton. Subsequently, the types of N doped in carbon materials can be studied by high-resolution N1s spectra. As shown in Figure 2c, the N1s spectra can be fitted well with 398.4 eV, 399.2 eV, and 402.2 eV, which represent pyridinic-N, pyrrolic-N, and graphitic-N, respectively. It should be noted that the pyridinic-N is the first place where carbon-based ORR catalyst absorbs O_2 and starts the ORR [37–40]. Meanwhile, the graphitic-N can become the electron-donor and facilitate charge mobility. As shown in Figure 2c and Figure S2, the CoNC material has a higher proportion of pyridinic-N and graphitic-N in N species than that of the NC material, which could further suggest that the CoNC is able to provide more efficient ORR performance than the NC.

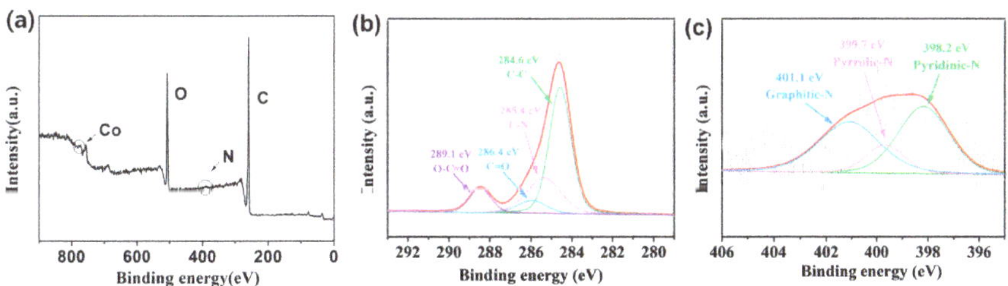

Figure 2. X-ray photoelectron spectroscopy (XPS) survey spectrum of CoNC (**a**), C1s and N1s XPS spectrum of CoNC (**b**,**c**).

2.2. Electrocatalytic Characteristics and Active Sites

The electrocatalytic activities of the four as-prepared samples were first investigated via cyclic voltammetry (CV) tests performed via an electrochemical workstation (CHI 750e) with three-electrode systems. As shown in Figure 3a and Figure S3, NC and CoNC show obvious redox peaks located at 0.69 V and 0.81 V vs. RHE at O_2-saturated KOH solution, and no obvious redox peak can be observed in CV curves at Ar-saturated KOH solution. This indicates that these redox peaks are formed by the ORR process. The CoNC catalyst possesses the positive ORR redox peak, which is comparable to the redox peak (0.82 V vs. RHE) shown in Pt/C (Figure 3b).

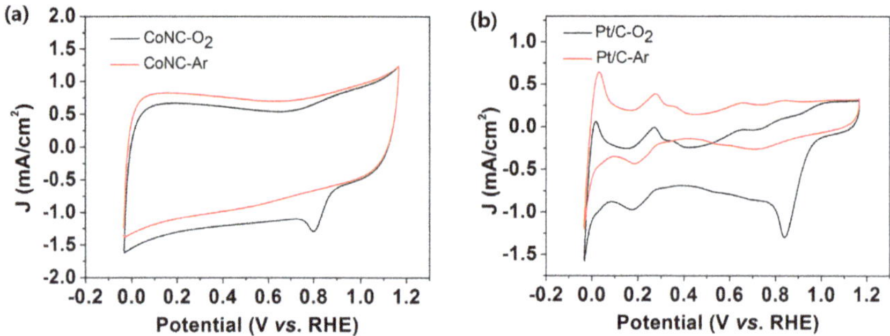

Figure 3. Cyclic voltammetry (CV) curves of CoNC (**a**) and Pt/C (**b**) in O_2- and Ar-saturated 0.1 M KOH solution.

To further study the catalytic activities and ORR kinetics of the as-prepared catalysts, linear sweep voltammetry (LSV) tests were characterized with an rotating disk electrode (RDE) electrode. As shown in corresponding LSV curves (Figure 4a,b, the bar plot corresponds to current density and the line corresponds to onset potential), CoNC shows the higher positive onset potential (0.86V vs. RHE) and limiting current density (4.91 mA cm^{-2}) than the PC (0.78 V, 4.33 mA cm^{-2}) and NC (0.84 V, 4.32 mA cm^{-2}) catalysts. Furthermore, these properties of CoNC are comparable to the benchmark catalyst of Pt/C, for which the onset potential and limiting current density are shown as 0.9 V and 5.67 mA cm^{-2}, respectively.

As numerous works reported, ORR has two major pathways to reduce oxygen into water—the two-electron pathway and the four-electron pathway, respectively. In the four-electron pathway, oxygen is directly reduced to water and is able to provide more positive potential during the ORR process, whereas the two-electron pathway first reduces O_2 to H_2O_2 then further reduces it to water. Thus, in order to investigate the catalytic pathway for CoNC, the ORR kinetics of CoNC were studied with RDE at different rotating speeds ranging from 100 rpm to 2500 rpm (Figure 4c). As shown in Figure 4d, LSV curves combined with the Koutecky-Lecich (K-L) equation are able to obtain several K-L plots at different potentials, which exhibits good parallelism and linearity, suggesting a first-order reaction toward ORR. The calculated electron transfer number is 3.73, which closed to the theoretical value 4, and suggests the four-electron dominant pathway during ORR. Meanwhile, Pt/C also shows the four-electron dominant pathway as electron transfer number (n = 4.03) calculated from K-L plots (Figure S4). For further study in ORR kinetic for CoNC, the rotating ring-disk electrode (RRDE) was used for testing the H_2O_2 productivity and electron transfer numbers while catalyzing ORR. As shown in Figure S5, CoNC produces a low proportion of H_2O_2 (less than 11.8%) and high electron transfer numbers (about 3.84) when ORR is carried out, while Pt/C shows 5.4% of H_2O_2 productivity and a 4.02 electron transfer number. These results elaborate that CoNC possesses the remarkable ORR catalytic activities in alkaline media and has the potential to be applied in metal-air batteries. The long-time durability tests were carried out by chronoamperometry measurement at 900 rpm in O_2-saturated 0.1 M KOH. As shown in Figure S6, after a 15000 s long-time durability test, the CoNC is still able to provide 94.82% of initial current density, while the Pt/C can only maintain 76.64% of initial current density. This indicates that CoNC holds excellent long-term durability, which further benefits the stability of batteries assembled with CoNC.

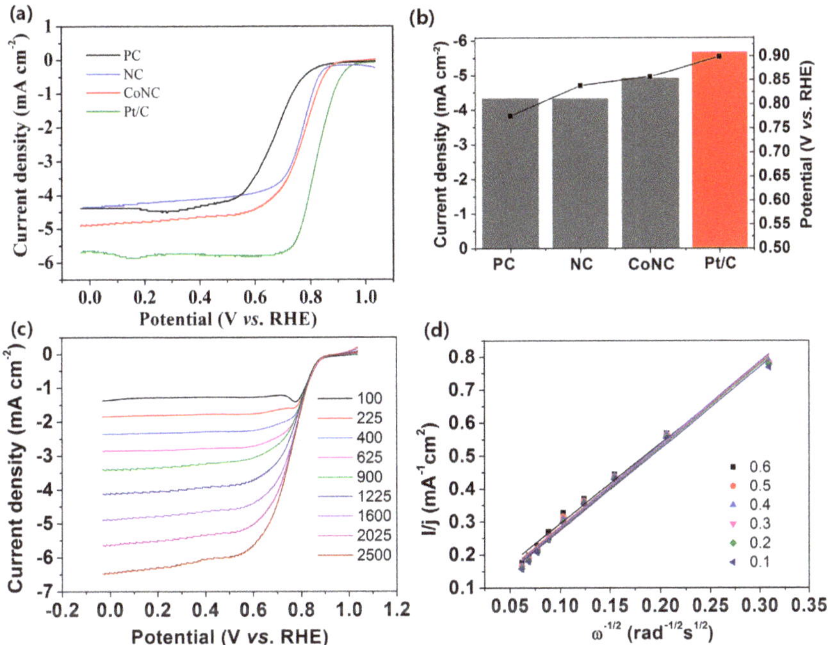

Figure 4. The linear sweep voltammetry (LSV) curves of PC, NC, CoNC and Pt/C catalysts at 1600 rpm (**a**).The onset potentials and limiting current density of catalysts (**b**).The LSV curves of CoNC catalyst at the rotation speeds of 100, 225, 400, 625, 900, 1225, 1600, 2025 and 2500 rpm (**c**). The corresponding Koutecky-Levich (K-L) plots (**d**).

Subsequently, the home-built Al-air batteries were assembled with CoNC and Pt/C in order to demonstrate the practical application performance of catalysts. As shown in Figure 5a, the polarization curves of Al-air batteries were first used for estimating the capability of Al-air batteries. The corresponding power density calculated from the polarization curve of CoNC shows that the maximum power density it can provide is 32.24 mW cm^{-2}, while the maximum power density for Al-air batteries assembled with Pt/C is 43.70 mW cm^{-2}. The electrochemical impedance spectroscopy (EIS) was used to analyze the resistance in Al-air batteries. As calculated by the Nyquist plots in Figure S7, the internal resistance of the Al-air battery with CoNC is 3.46 Ω, which is comparable to that of the battery with 20 wt% Pt/C (3.29 Ω). In order to assess the long-term working property of the battery, the galvanostatic discharge performance was tested in a constant current of 20 mA cm^{-2} (Figure 5b). After a 5000 s test, the voltages of batteries with CoNC and Pt/C decreased 0.04 V and 0.21 V, respectively, and the voltages of batteries with CoNC remained at the constant discharge voltage of 1.17 V at 20 mA cm^{-2}. The superior durability of batteries assembled with CoNC can be attributed to the excellent catalytic durability of CoNC.

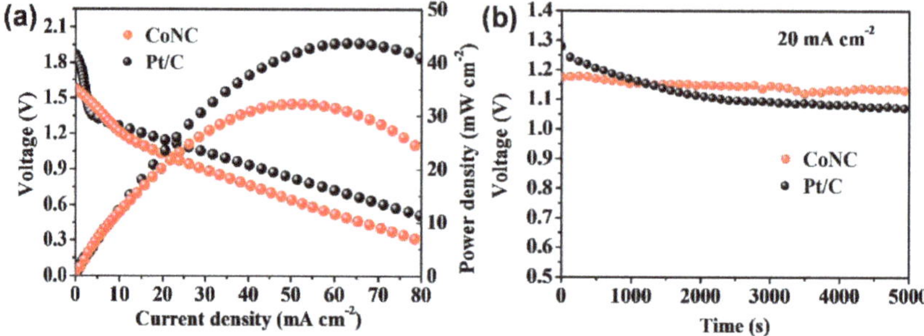

Figure 5. The polarization curves and corresponding power densities of the Al-air batteries with two catalysts (**a**). The galvanostatic discharge curves of Al-air batteries with two catalysts at the constant current density of 20 mA cm^{-2} (**b**).

3. Experimental Section

3.1. Preparation of Co and N Co-Doped Biocarbon

The Co and N co-doped biocarbon is synthesized by a one-step pyrolysis method. Specifically, a mixture of chitin (2 g, MYM biological technology company limited, Nanjing, China), $CoCl_2 \cdot H_2O$ (0.6 g, Fuchen chemical reagents factory, Tianjin, China) and urea (2 g, Beijing Chemical Works, Beijing, China) were ground thoroughly and dissolved into 40 mL of deionized water. Then the mixture was stirred overnight with a string speed of 600 rpm to form a homogeneous solution that was then dried at 60 °C for 48 h. Subsequently, the resulting sample was pyrolyzed at 800 °C for 1 h at a heating rate of 5 °C min^{-1} (Sager SG-GS1700). The resultant powder, N and Co co-doped biocarbon material, was denoted as CoNC. As controls, directly carbonized chitin (PC) and N-doped bio-carbon (NC) without any Co, were also prepared by the same method. For comparison, 20 wt% Pt/C was prepared and the loading for Pt was about 20 μg cm^{-2}.

3.2. Physicochemical Characterization

The morphologies of as-prepared catalysts were studied by scanning electron microscopy (SEM) with a JEOL JSM-7500F field emission scanning electron microscope and transmission electron microscope (TEM) with a JEOL-2010 electron microscope operating at 200 kV. X-ray photoelectron spectroscopy (XPS) was performed on ESCALAB 250Xi (Thermo Scientific, Shanghai, China). A laser microscope system (B&W TEK BTC 162E) with an exciting laser of 532 nm was applied to study the Raman shifts. Then, the specific surface area and pore size were studied via automated gas sorption analyzer (ASAP 2020M-Physisoprtion Analyzer, Changchun, China) at 77 K. The Brunauer–Emmett–Teller method was used to calculate the specific surface area and the pore size.

3.3. Electrochemical Measurements

The electrochemical measurements were carried on a CHI 750E electrochemical workstation system with a typical three-electrode cell in 0.1 M KOH solution at room temperature. Glassy carbon electrode, Ag/AgCl, and Pt wire were used as a working electrode, reference electrode, and counter electrode, respectively. The catalyst inks were prepared by ultrasonically dispersing 5 mg of as-prepared catalysts, 5 mg of carbon black, and 50 μL of Nafion in 1 mL of ethanol. Then, 20 μL well-dispersed catalyst inks were loaded on a working electrode (5.5 mm diameter), leading to a 100 μg cm^{-2} catalyst loading. The as-prepared electrodes were dried completely at room temperature before the electrochemical tests. Cyclic voltammetry (CV) was measured in O_2- and Ar-saturated 0.1 M KOH electrolyte with a scan rate of 10 mV s^{-1}. For the ORR kinetics study, the linear sweep voltammetry (LSV) tests were performed at rotating rates from 100 to 2500 in O_2-saturated electrolyte with a scan speed of 10 mV s^{-1}. The long-time durability of catalysts was tested in O_2-saturated 0.1 M KOH by chronoamperometry.

The electrons transfer number during ORR was analyzed from RDE measurements by using the following Koutecky-Lecich (K-L) equations (Equations (1) and (2)):

$$j^{-1} = j_K{}^{-1} + j_L{}^{-1} = j_K{}^{-1} + B^{-1}\omega^{-1/2} \quad (1)$$

$$\mathrm{B} = 0.2\mathrm{FC_0}(\mathrm{D_0})^{2/3}v^{-1/6} \quad (2)$$

where j_L and j_K are the limiting current for the electrode reaction of reactive species by the diffusion-controlled process and the kinetic current, respectively. F (96486.4 C mol^{-1}) is the Faraday constant, n is the electron transfer number, C_0* (1.2×10^{-6} mol cm^{-3}) is the concentration of O_2 in 0.1 M KOH solution, D_0 (1.9×10^{-5} cm^2 s^{-1}) is the diffusion coefficient of O_2 in 0.1 M KOH, ω (rad s^{-1}) is rotation rate and v (0.01 cm^2 s^{-1}) is the kinematic viscosity of the electrolyte [41,42].

3.4. Al-Air Battery Tests

The measurements of Al-air batteries were tested via home-built electrochemical cells. Al plate (99% purity) and 6 M KOH aqueous solution were used as the anode and electrolyte, respectively. A cathode was prepared by catalyst layer, carbon cloth (middle layer), carbon-based layer, and the diffusion layer. The mixture of carbon black and 40% Polytetrafluoroethylene (PTFE) was coated onto carbon cloth to form a carbon-based layer that prevents penetration of PTFE in the following process. A diffusion layer was prepared by applying a 60% PTFE layer coating on the carbon-based layer and heating it at 370 °C for 20 min. Subsequently, a catalyst layer was prepared by loading a mixture of CoNC, 5% Nafion, and ethanol on the other side of the carbon cloth (25 mg cm^{-2}). For comparison, the Al-air battery that used commercial Pt/C as an air cathode was fabricated via the same procedures. The polarization curves and long-term discharge performances were tested by electrochemical workstation (CHI 750E) under air atmosphere.

4. Conclusions

In conclusion, a low-cost method was developed to synthesize efficient ORR catalyst derived from chitin. The biomass-derived electrocatalyst (CoNC) possesses high specific surface area, which contributes to enhanced ORR performance, and shows good ORR catalytic activities during ORR in alkaline media. In addition, Al-air batteries equipped with CoNC electrocatalyst show good power density and excellent long-term discharging durability, which is attributed to the efficient catalytic activities of CoNC. This work paves a new way for mass production of low-cost ORR catalyst and the utilization of biomass product.

Supplementary Materials: The following are available online at http://www.mdpi.com/2073-4344/9/11/954/s1, Figure S1: The SEM and TEM images of NC, Figure S2: N1s XPS spectrum of NC, Figure S3: CV curves of NC catalysts, Figure S4: The LSV curves of Pt/C catalyst at the rotation speeds of 100, 225, 400, 625, 900, 1225, 1600, 2025 and 2500 rpm (a). The corresponding Koutecky-Levich (K-L) plots (b), Figure S5: The H_2O_2 yield and n value of CoNC and Pt/C, Figure S6: The i-t chronoamperometric curves of CoNC and Pt/C for 15000 s in an O_2-saturated 0.1 M KOH solution, Figure S7: Nyquist plots of CoNC and commercial Pt/C from 100 KHz to 0.01 Hz.

Author Contributions: G.L. and S.Y. designed experiments; M.W. and J.M. carried out experiments; M.W. and Z.C. analyzed sequencing data; M.W. and H.Y. wrote the manuscript.

Funding: This work was funded by the National Natural Science Foundation of China (grant number: 51605188, 51605187), Department of Education of Jilin Province (grant number: JJKH20180093KJ, JJKH20181163KJ), and Jilin Provincial Science & Technology Department (Grant Number: 20180201004GX).

Conflicts of Interest: The author declares that there is no conflict of interests regarding the publication of this article.

References

1. Liu, Z.N.; Li, Z.Y.; Ma, J.; Dong, X.; Ku, W.; Wang, M.; Sun, H.; Liang, S.; Lu, G.L. Nitrogen and cobalt-doped porous biocarbon materials derived from corn stover as efficient electrocatalysts for aluminum-air batteries. *Energy* **2018**, *162*, 453–459. [CrossRef]
2. Guo, Z.Y.; Li, C.; Liu, J.Y.; Wang, Y.G.; Xia, Y.Y. A Long-Life Lithium-Air Battery in Ambient Air with a Polymer Electrolyte Containing a Redox Mediator. *Angew. Chem. Int. Ed.* **2017**, *56*, 7505–7509. [CrossRef] [PubMed]
3. Pan, J.; Xu, Y.Y.; Yang, H.; Dong, Z.; Liu, H.; Xia, B.Y. Advanced Architectures and Relatives of Air Electrodes in Zn-Air Batteries. *Adv. Sci. (Weinh.)* **2018**, *5*, 1700691. [CrossRef] [PubMed]
4. Ryu, J.; Park, M.; Cho, J. Advanced Technologies for High-Energy Aluminum-Air Batteries. *Adv. Mater.* **2019**, *31*, 1804784. [CrossRef] [PubMed]
5. Li, J.Z.; Chen, M.J.; Cullen, D.A.; Hwang, S.; Wang, M.Y.; Li, B.Y.; Liu, K.X.; Karakalos, S.; Lucero, M.; Zhang, H.G.; et al. Atomically dispersed manganese catalysts for oxygen reduction in proton-exchange membrane fuel cells. *Nat. Catal.* **2018**, *1*, 935–945. [CrossRef]
6. Yang, D.J.; Zhang, L.J.; Yan, X.C.; Yao, X.D. Recent Progress in Oxygen Electrocatalysts for Zinc-Air Batteries. *Small Methods* **2017**, *1*, 1700209. [CrossRef]
7. Liu, Y.; Liu, Z.; Liu, H.; Liao, M. Novel Porous Nitrogen Doped Graphene/Carbon Black Composites as Efficient Oxygen Reduction Reaction Electrocatalyst for Power Generation in Microbial Fuel Cell. *Nanomaterials* **2019**, *9*, 836. [CrossRef]
8. Liu, J.; Jiao, M.G.; Lu, L.L.; Barkholtz, H.M.; Li, Y.P.; Wang, Y.; Jiang, L.H.; Wu, Z.J.; Liu, D.J.; Zhuang, L.; et al. High performance platinum single atom electrocatalyst for oxygen reduction reaction. *Nat. Commun.* **2017**, *8*, 15938. [CrossRef]
9. Liu, H.Y.; Qin, J.Q.; Zhao, S.Q.; Gao, Z.M.; Fu, Q.; Song, Y.J. Two-dimensional circular platinum nanodendrites toward efficient oxygen reduction reaction and methanol oxidation reaction. *Electrochem. Commun.* **2019**, *98*, 53–57. [CrossRef]
10. Li, Y.; Xia, X.H. Study on the Mechanism and Kinetics of Oxygen Reduction Reaction on 3D Porous Platinum Film Constructed Using Colloidal Crystal Template. *J. Nanosci. Nanotech.* **2016**, *16*, 12388–12393. [CrossRef]
11. Lange, K.; Schulz-Ruhtenberg, M.; Caro, J. Platinum Electrodes for Oxygen Reduction Catalysis Designed by Ultrashort Pulse Laser Structuring. *Chemelectrochem* **2017**, *4*, 570–576. [CrossRef]
12. Oh, T.; Kim, K.; Kim, J. Controllable active sites and facile synthesis of cobalt nanoparticle embedded in nitrogen and sulfur co-doped carbon nanotubes as efficient bifunctional electrocatalysts for oxygen reduction and evolution reactions. *J. Energy Chem.* **2019**, *38*, 60–67. [CrossRef]
13. Kang, Y.S.; Choi, D.; Park, H.-Y.; Yoo, S.J. Tuning the surface structure of PtCo nanocatalysts with high activity and stability toward oxygen reduction. *J. Ind. Eng. Chem.* **2019**, *78*, 448–454. [CrossRef]
14. Kim, S.; Kato, S.; Ishizaki, T.; Li, O.L.; Kang, J. Transition Metal (Fe, Co, Ni) Nanoparticles on Selective Amino-N-Doped Carbon as High-Performance Oxygen Reduction Reaction Electrocatalyst. *Nanomaterials* **2019**, *9*, 742. [CrossRef] [PubMed]
15. Zhang, B.; Xiao, C.H.; Xiang, Y.; Dong, B.T.; Ding, S.J.; Tang, Y.H. Nitrogen-Doped Graphene Quantum Dots Anchored on Thermally Reduced Graphene Oxide as an Electrocatalyst for the Oxygen Reduction Reaction. *Chemelectrochem* **2016**, *3*, 864–870. [CrossRef]
16. Su, C.Y.; Cheng, H.; Li, W.; Liu, Z.Q.; Li, N.; Hou, Z.F.; Bai, F.Q.; Zhang, H.X.; Ma, T.Y. Atomic Modulation of FeCo-Nitrogen-Carbon Bifunctional Oxygen Electrodes for Rechargeable and Flexible All-Solid-State Zinc-Air Battery. *Adv. Energy Mater.* **2017**, *7*, 1602242. [CrossRef]
17. Zhu, J.W.; Li, W.Q.; Li, S.H.; Zhang, J.; Zhou, H.; Zhang, C.T.; Zhang, J.A.; Mu, S.C. Defective N/S-Codoped 3D Cheese-Like Porous Carbon Nanomaterial toward Efficient Oxygen Reduction and Zn-Air Batteries. *Small* **2018**, *14*, 1800563. [CrossRef]
18. Ma, Z.; Wang, K.X.; Qiu, Y.F.; Liu, X.Z.; Cao, C.Y.; Feng, Y.J.; Hu, P.A. Nitrogen and sulfur co-doped porous carbon derived from bio-waste as a promising electrocatalyst for zinc-air battery. *Energy* **2018**, *143*, 43–55. [CrossRef]
19. Wang, G.H.; Deng, Y.J.; Yu, J.N.; Zheng, L.; Du, L.; Song, H.Y.; Liao, S.J. From Chlorella to Nestlike Framework Constructed with Doped Carbon Nanotubes: A Biomass-Derived, High-Performance, Bifunctional Oxygen Reduction/Evolution Catalyst. *ACS Appl. Mater. Interface* **2017**, *9*, 32168–32178. [CrossRef]

20. Zhang, Z.P.; Gao, X.J.; Dou, M.L.; Ji, J.; Wang, F. Biomass Derived N-Doped Porous Carbon Supported Single Fe Atoms as Superior Electrocatalysts for Oxygen Reduction. *Small* **2017**, *13*, 1604290. [CrossRef]
21. Xu, L.N.; Fan, H.; Huang, L.X.; Xia, J.L.; Li, S.H.; Li, M.; Ding, H.Y.; Huang, K. Chrysanthemum-derived N and S co-doped porous carbon for efficient oxygen reduction reaction and aluminum-air battery. *Electrochim. Acta* **2017**, *239*, 1–9. [CrossRef]
22. Liu, L.; Yang, X.F.; Ma, N.; Liu, H.T.; Xia, Y.Z.; Chen, C.M.; Yang, D.J.; Yao, X.D. Scalable and Cost-Effective Synthesis of Highly Efficient Fe2N-Based Oxygen Reduction Catalyst Derived from Seaweed Biomass. *Small* **2016**, *12*, 1295–1301. [CrossRef] [PubMed]
23. Lu, G.; Zhu, Y.; Lu, L.; Xu, K.; Wang, H.; Jin, Y.; Ren, Z.J.; Liu, Z.; Zhang, W. Iron-rich nanoparticle encapsulated, nitrogen doped porous carbon materials as efficient cathode electrocatalyst for microbial fuel cells. *J. Power Sources* **2016**, *315*, 302–307. [CrossRef]
24. Lu, G.; Zhu, Y.; Xu, K.; Jin, Y.; Ren, Z.J.; Liu, Z.; Zhang, W. Metallated porphyrin based porous organic polymers as efficient electrocatalysts. *Nanoscale* **2015**, *7*, 18271–18277. [CrossRef]
25. Yan, L.; Yu, J.; Houston, J.; Flores, N.; Luo, H. Biomass Derived Porous Nitrogen doped Carbon for Electrochemical Devices. *Green Energy Environ.* **2017**, *2*, 84–99. [CrossRef]
26. Li, Y.H.; Liu, D.B.; Gan, J.; Duan, X.Z.; Zang, K.T.; Ronning, M.; Song, L.; Luo, J.; Chen, D. Sustainable and Atomically Dispersed Iron Electrocatalysts Derived from Nitrogen- and Phosphorus-Modified Woody Biomass for Efficient Oxygen Reduction. *Adv. Mater. Interfaces* **2019**, *6*, 1801623. [CrossRef]
27. Wu, D.Y.; Zhu, C.; Shi, Y.T.; Jing, H.Y.; Hu, J.W.; Song, X.D.; Si, D.H.; Liang, S.X.; Hao, C. Biomass-Derived Multilayer-Graphene-Encapsulated Cobalt Nanoparticles as Efficient Electrocatalyst for Versatile Renewable Energy Applications. *ACS Sustain. Chem. Eng.* **2019**, *7*, 1137–1145. [CrossRef]
28. Wei, Q.L.; Yang, X.H.; Zhang, G.X.; Wang, D.N.; Zuin, L.; Banham, D.; Yang, L.J.; Ye, S.Y.; Wang, Y.L.; Mohamedi, M.; et al. An active and robust Si-Fe/N/C catalyst derived from waste reed for oxygen reduction. *Appl. Catal. B Environ.* **2018**, *237*, 85–93. [CrossRef]
29. Yuan, H.; Deng, L.; Cai, X.; Zhou, S.; Chen, Y.; Yuan, Y. Nitrogen-doped carbon sheets derived from chitin as non-metal bifunctional electrocatalysts for oxygen reduction and evolution. *RSC Adv.* **2015**, *5*, 56121–56129. [CrossRef]
30. Bo, W.; Li, S.; Wu, X.; Liu, J.; Jing, C. Biomass chitin-derived honeycomb-like nitrogen-doped carbon/graphene nanosheet networks for applications in efficient oxygen reduction and robust lithium storage. *J. Mater. Chem. A* **2016**, *4*. [CrossRef]
31. Chen, J.; Qiu, L.; Li, Z.; Gao, G.; Zhong, W.; Zhang, P.; Gong, Y.; Deng, L. Chitin-derived porous carbon loaded with Co, N and S with enhanced performance towards electrocatalytic oxygen reduction, oxygen evolution, and hydrogen evolution reactions. *Electrochim. Acta* **2019**, *304*, 350–359. [CrossRef]
32. Borghei, M.; Laocharoen, N.; Kibena-Poldsepp, E.; Johansson, L.S.; Campbell, J.; Kauppinen, E.; Tammeveski, K.; Rojas, O.J. Porous N,P-doped carbon from coconut shells with high electrocatalytic activity for oxygen reduction: Alternative to Pt-C for alkaline fuel cells. *Appl. Catal. B Environ.* **2017**, *204*, 394–402. [CrossRef]
33. Jiang, Z.Q.; Zhao, X.S.; Tian, X.N.; Luo, L.J.; Fang, J.H.; Gao, H.Q.; Jiang, Z.J. Hydrothermal Synthesis of Boron and Nitrogen Codoped Hollow Graphene Microspheres with Enhanced Electrocatalytic Activity for Oxygen Reduction Reaction. *ACS Appl. Mater. Interface* **2015**, *7*, 19398–19407. [CrossRef] [PubMed]
34. Meng, F.L.; Zhong, H.X.; Bao, D.; Yan, J.M.; Zhang, X.B. In Situ Coupling of Strung Co4N and Intertwined N-C Fibers toward Free-Standing Bifunctional Cathode for Robust, Efficient, and Flexible Zn Air-Batteries. *J. Am. Chem. Soc.* **2016**, *138*, 10226–10231. [CrossRef] [PubMed]
35. Chen, Z.; Gao, X.; Wei, X.; Wang, X.; Li, Y.; Wu, T.; Guo, J.; Gu, Q.; Wu, W.D.; Chen, X.D. Directly anchoring Fe 3 C nanoclusters and FeN x sites in ordered mesoporous nitrogen-doped graphitic carbons to boost electrocatalytic oxygen reduction. *Carbon* **2017**, *121*, 143–153. [CrossRef]
36. Li, C.L.; Wu, M.C.; Liu, R. High-performance bifunctional oxygen electrocatalysts for zinc-air batteries over mesoporous Fe/Co-N-C nanofibers with embedding FeCo alloy nanoparticles. *Appl. Catal. B Environ.* **2019**, *244*, 150–158. [CrossRef]
37. Liu, Z.; Li, Z.; Tian, S.; Wang, M.; Sun, H.; Liang, S.; Chang, Z.; Lu, G. Conversion of peanut biomass into electrocatalysts with vitamin B12 for oxygen reduction reaction in Zn-air battery. *Int. J. Hydrogen Energy* **2019**, *44*, 11788–11796. [CrossRef]

38. Wang, Y.; Zhu, M.; Wang, G.; Dai, B.; Yu, F.; Tian, Z.; Guo, X. Enhanced Oxygen Reduction Reaction by In Situ Anchoring Fe_2N Nanoparticles on Nitrogen-Doped Pomelo Peel-Derived Carbon. *Nanomaterials* **2017**, *7*, 404. [CrossRef]
39. Lu, G.; Li, Z.; Fan, W.; Wang, M.; Yang, S.; Li, J.; Chang, Z.; Sun, H.; Liang, S.; Liu, Z. Sponge-like N-doped carbon materials with Co-based nanoparticles derived from biomass as highly efficient electrocatalysts for the oxygen reduction reaction in alkaline media. *RSC Adv.* **2019**, *9*, 4843–4848. [CrossRef]
40. Lu, J.; Zeng, Y.; Ma, X.; Wang, H.; Gao, L.; Zhong, H.; Meng, Q. Cobalt Nanoparticles Embedded into N-Doped Carbon from Metal Organic Frameworks as Highly Active Electrocatalyst for Oxygen Evolution Reaction. *Polymers* **2019**, *11*, 828. [CrossRef]
41. Stosevski, I.; Krstic, J.; Milikic, J.; Sljukic, B.; Kacarevic-Popovic, Z.; Mentus, S.; Miljanic, S. Radiolitically synthesized nano Ag/C catalysts for oxygen reduction and borohydride oxidation reactions in alkaline media, for potential applications in fuel cells. *Energy* **2016**, *101*, 79–90. [CrossRef]
42. Zhang, J.W.; Xu, D.; Wang, C.C.; Guo, J.N.; Yan, F. Rational Design of Fe1-xS/Fe3O4/Nitrogen and Sulfur-Doped Porous Carbon with Enhanced Oxygen Reduction Reaction Catalytic Activity. *Adv. Mater. Interfaces* **2018**, *5*, 1701461.

© 2019 by the authors. Licensee MDPI, Basel, Switzerland. This article is an open access article distributed under the terms and conditions of the Creative Commons Attribution (CC BY) license (http://creativecommons.org/licenses/by/4.0/).

Article

Pt-Co_3O_4 Superstructures by One-Pot Reduction/Precipitation in Bicontinuous Microemulsion for Electrocatalytic Oxygen Evolution Reaction

Elijah T. Adesuji [1,2], Esther Guardado-Villegas [1], Keyla M. Fuentes [1], Margarita Sánchez-Domínguez [1,*] and Marcelo Videa [2,*]

1 Group of Colloidal and Interfacial Chemistry Applied to Nanomaterials and Formulations, Centro de Investigación en Materiales Avanzados, S. C. (CIMAV), Unidad Monterrey, Alianza Norte 202, Parque de Investigación e Innovación Tecnológica, Apodaca 66628, Mexico; elijah.adesuji@cimav.edu.mx (E.T.A.); estherguardadov@gmail.com (E.G.-V.); keyla.fuentes@cimav.edu.mx (K.M.F.)

2 Department of Chemistry and Nanotechnology, School of Engineering and Sciences, Tecnológico de Monterrey, Ave. Eugenio Garza Sada 2501, Monterrey 64849, Mexico

* Correspondence: margarita.sanchez@cimav.edu.mx (M.S.-D.); mvidea@itesm.mx (M.V.)

Received: 11 October 2020; Accepted: 3 November 2020; Published: 12 November 2020

Abstract: Bicontinuous microemulsions (BCME) were used to synthesize hierarchical superstructures (HSs) of Pt-Co_3O_4 by reduction/precipitation. BCMEs possess water and oil nanochannels, and therefore, both hydrophilic and lipophilic precursors can be used. Thus, PtAq-CoAq, PtAq-CoOi, PtOi-CoAq and PtOi-CoOi were prepared (where Aq and Oi stand for the precursor present in aqueous or oily phase, respectively). The characterization of the Pt-Co_3O_4-HS confirmed the formation of metallic Pt and Co_3O_4 whose composition and morphology are controlled by the initial pH and precursor combination, determining the presence of the reducing/precipitant species in the reaction media. The electrocatalytic activity of the Pt-Co_3O_4-HSs for oxygen evolution reaction (OER) was investigated using linear sweep voltammetry in 0.1 M KOH and compared with Pt-HS. The lowest onset overpotentials for Pt-Co_3O_4-Hs were achieved with PtOi-CoOi (1.46 V vs. RHE), while the lowest overpotential at a current density of 10 mA cm^{-2} (η_{10}) was obtained for the PtAq-CoAq (381 mV). Tafel slopes were 102, 89, 157 and 92 mV dec^{-1}, for PtAq-CoAq, PtAq-CoOi, PtOi-CoAq and PtOi-CoOi, respectively. The Pt-Co_3O_4-HSs showed a better performance than Pt-HS. Our work shows that the properties and performance of metal–metal oxide HSs obtained in BCMEs depend on the phases in which the precursors are present.

Keywords: superstructures; bicontinuous microemulsion; oxygen evolution reaction; metal–metal oxides

---

1. Introduction

Electrocatalytic reactions involving HER (hydrogen evolution reaction) and OER (oxygen evolution reaction) are required in the production of hydrogen and oxygen by water electrolysis. These reactions, relevant to renewable energy technologies, require the lowering of onset potential and increasing the reaction kinetics to make the energy balance efficient [1–3].

$$2H_2O + 2e^- \rightarrow 2OH^- + H_2 \qquad \text{Hydrogen evolution reaction} \tag{1}$$

$$4OH^- \rightarrow 2H_2O + 4e^- + O_2 \qquad \text{Oxygen evolution reaction} \tag{2}$$

Generally, anodic OER is characterized by a high onset potential, and it exhibits slow kinetics because the reaction requires four-electron transfer and oxygen-oxygen bond formation [4]. These characteristics hamper large-scale water splitting technologies. Therefore, to overcome large onset potential and sluggish anodic OER, the use of an appropriate catalyst is important. RuO_2 and IrO_2 are widely used as OER catalysts in achieving water electrolysis. However, they are costly, scarce and have unstable catalytic performance in alkaline media [5,6]. These factors have prompted various research efforts geared towards exploring readily available, highly active, cost-effective and efficient catalysts that can lower the onset potential, have good stability and increase the reaction kinetics [7–9]. Studies have shown that Co_3O_4-based nanomaterials are affordable, abundant in nature, resistant to corrosion and possess enhanced catalytic performance towards OER in alkaline media [3,10,11].

Spectroscopic methods such as ex situ electron paramagnetic resonance spectroscopy [12] and in situ Raman spectroscopy have established the formation of Co^{4+} species in Co_3O_4-based nanomaterials as the active centers for OER. The EPR spectroscopic studies of Cobalt-phosphate catalyst evidenced the generation and population rise in active Co^{4+} species during water oxidation. An increase from 3 to 7% in Co^{4+} population improved the water oxidation by more than 2 orders of magnitude, with a change in the deposition potential from 1.14 to 1.34 V vs. NHE [13]. The in situ Raman spectra of a CoO_x monolayers deposited on gold acquired during linear sweep voltammetry from 0 to 1.0 V vs. Hg|HgO in 0.1 M KOH, revealed the oxidation of the Co^{2+} and Co^{3+} species to Co^{4+} by increasing the potential [14]. Therefore, a strategy to increase the population of Co^{4+} centers on the surface of cobalt oxide will increase its performance as an OER catalyst. The introduction of highly electronegative transition metals by electrochemical or chemical methods, as support in a metal–metal oxide composites has been successful in achieving this.

In recent literature, metal–cobalt oxide nanomaterials have shown better OER performances than either noble metal or Co_3O_4 nanomaterials alone [10,15,16]. During OER, the presence of noble metals or highly electronegative metal ions promotes the oxidation of Co^{3+} to Co^{4+} through electron-withdrawing inductive effect [10,14,17]. The galvanostatic deposition of cobalt oxide on Au support by Yeo and Bell. 2011 resulted in improved OER performances. The turnover frequencies at an overpotential of 351 mV for 75 μC of cobalt oxide deposited on Au, is 1.24 s^{-1} compared to 0.0071 s^{-1} for Au without cobalt oxide deposit [14]. The successful synthesis of Au@Co_3O_4 nanocrystal (NC) catalysts using hydrothermal method has been reported. The draining of electrons from Co_3O_4 by Au and the stronger binding to oxygen by Co_3O_4 shell of Au@Co_3O_4 nanocrystals increases the Co^{4+} population, resulting in enhanced OER activity. At a fixed potential of 1.58 V vs. RHE, Au@Co_3O_4 exhibited a current density of 2.84 mA cm^{-2}, which is 7 times as high as Co_3O_4 having 0.42 mA cm^{-2} and 55 times that of Au (0.05 mA cm^{-2}) [17]. Qu et al., 2017 demonstrated the ability of Pd-Co_3O_4 nanostructures to generate high amounts of Co^{4+} cations. Superior electrochemical performance deemed by the onset potential and Tafel slope for Pd-Co_3O_4 (1.388 V vs. RHE and 60.7 mV dec^{-1}) in comparison with Co_3O_4/C (1.481 V vs. RHE and 96.1 mV dec^{-1}) were reported. This was attributed to the synergistic effect between Pd and Co_3O_4 and the high surface area of the 3D ordered mesoporous structure, which enhanced the charge transfer between the solid/electrolyte interface. High OER activity has been reported for Au-Co_3O_4/C electrocatalyst compared to Co_3O_4 nanomaterial. A value of 1.507 V vs. RHE for the onset potential of Au-Co_3O_4/C was recorded while Au/C and Co_3O_4 had values of 1.568 V vs. RHE and 1.518 V vs. RHE, respectively. This is suggested, once again, to be due to the charge transfer from the cobalt oxide to electronegative Au, resulting in an increase in the formation of Co^{4+} cations. The Au nanoparticles with an average diameter of 7 nm were well dispersed in Co_3O_4 [10]. Recently, the synthesis of noble metal particles (Pt, Pd, Au) embedded within mesoporous Co_3O_4 resulted in materials exhibiting superior activities and excellent stability in alkaline medium when compared with either the metal catalysts/C (Pt, Pd, Au) or Co_3O_4/C [16]. An onset potential of 1.481 V vs. RHE was recorded on Co_3O_4/C electrocatalyst, while 1.582 V, 1.538 V and 1.618 V vs. RHE for Pt/C, Pd/C and Au/C electrodes were reported, respectively. The metal–cobalt oxide electrocatalysts of Pt-Co_3O_4, Pd-Co_3O_4 and Au-Co_3O_4 had onset potentials of 1.383V, 1.388 V and 1.395V vs. RHE, respectively. The OER kinetic measurements showed Tafel slopes of 57.55, 60.70 and 68.01 mV dec^{-1} for Pt-Co_3O_4, Pd-Co_3O_4 and Au-Co_3O_4 electrocatalyst, respectively. This showed that the introduction of metals into cobalt oxides promotes higher OER activity than that observed on either the metal or cobalt oxide nanomaterial alone.

Furthermore, it has been shown that oxygen vacancies in Co_3O_4-based nanomaterials would result in a higher Co^{2+}/Co^{3+} ratio, but conversely provide more active sites and lead to improved performance for OER. Xiao et al. 2020 studied the real active sites during OER process using operando characterization techniques such as EIS (electrochemical impedance spectroscopy), CV (cyclic voltammetry), XPS (X-ray photoelectron spectroscopy) and showed that oxygen vacancies promote the pre-oxidation of low-valence Co (Co^{2+}) as well as the formation of cobalt oxyhydroxide (Co^{III}-OOH) intermediate species for OER. The authors also showed that a high ratio of Co^{2+}/Co^{3+} plays a role in the rapid OH ion adsorption and deprotonation of CoOOH [18].

A wide range of noble metal–cobalt oxide nanomaterials used as OER electrocatalysts have been synthesized by the chemical reduction method [6,19], nanocasting using highly ordered mesoporous SiO_2 as a hard template [4,16,20–23], and the hydrothermal method [24]. Likewise, the synthesis of metal nanoparticles (mono and bimetallic), as well as porous materials, could be obtained in one-pot procedures by using microemulsions as confined reaction media and template [25–27].

Recently, we reported for the first time the synthesis of hierarchical metallic superstructures using bicontinuous microemulsions (BCME) [28]. A nano-meso-macrostructure of Pt electrocatalyst with a nanocoral morphology was obtained from a soft and facile synthesis approach. Since BCMEs comprise both water and oil interconnected nanochannels, it was possible to obtain platinum hierarchical superstructures (Pt-HSs) using both water-soluble and oil-soluble precursors; it was observed that the resultant morphology depends on the type of Pt precursor. The resultant materials showed an efficient and stable electrocatalytic performance for hydrogen evolution reaction. In the present work, we extend this strategy for the synthesis of a hybrid metal–metal oxide HSs material. To the best of our knowledge, synthesis of Pt-Co_3O_4 superstructures by bicontinuous microemulsion (BCME) has not been previously investigated. The synthesis of Pt-Co_3O_4 and Pt hierarchical superstructures with various characteristics and performance is achieved by choosing different combinations of Co and Pt metallic precursors, based on their solubility, and employing bicontinuous microemulsion as a soft template, for achieving unique hierarchical morphologies.

2. Results and Discussion

A one-step chemical reduction/precipitation process was used to synthesize the Pt-Co_3O_4-HSs. The structures of the nanomaterials obtained reflect the confinement imposed by the bicontinuous microemulsion which acted as an interconnected nanocage network. The SEM images reveal that the Pt-Co_3O_4-HSs, as well as the reference Pt-HS solid, are made up of hierarchical nanostructures with a unique tailored arrangement. Figure 1A,B show the morphology of the Pt-HS. An interconnected-needle-like structure resembling a nanocoral is observed for Pt-HS. The SEM images for PtAq-CoAq revealed a morphological arrangement similar to that of Pt-HS (Figure 1C,D). This seems to follow from the fact that the precursors for PtAq-CoAq and Pt-HS are dissolved in the same medium in which the reducing agent $NaBH_4$ was dissolved. Therefore, the resulting framework can be attributed to the soft template, morphology-directing nature of the water channels in the BCME and the presence of $NaBH_4$ in the same aqueous phase. Similar observation has been reported by Zhuang et al., where the authors attributed the formation of hierarchical microstructure to the use of CTAB micelles and a strong reducing agent dissolved in the same aqueous solution as the metal precursors [6]. In the reacting systems containing either or both metallic precursors dissolved in the organic phase; (PtAq-CoOi, PtOi-CoAq, and PtOi-CoOi), the SEM images (Figure 1E–J), show a different morphology from that of Pt-HS. In the case of PtAq-CoOi, nanoplatelet-like morphology was observed, and PtOi-CoOi, exhibited a fiber-like morphology, which is attributed to the nanocage property of the oily channels in our BCME and the restricted access of reducing agent through the interface. Also, it is expected that the oil channels to be wider [28] than the water channels, since there is a larger fraction of the oil phase in the BCME, which permits the formation of different shapes when one or both precursors are present in the oil phase. Figure 2A,C present the STEM images of PtAq-CoAq and PtOi-CoAq. Agglomeration of interconnected needle-like nanomaterial was obtained for PtAq-CoAq, while PtOi-CoAq showed small nanoparticle agglomerates without a defined shape. Figure 2B,D show that PtAq-CoOi and PtOi-CoOi are made up of nanoplatelet-like and fiber-like nanomaterials respectively, which can also observed from SEM.

The Energy-dispersive X-ray spectroscopy (EDS) analysis showed the presence of Pt, Co, and O in all the samples (Figure S1). The ICP-AES analysis results in Table 1 reveal that Pt-Co_3O_4-HSs synthesized using Pt aqueous precursor yielded higher amounts of Pt than the nominal Pt/Co mole ratios. This observation suggests that a higher Pt reduction was achieved for the aqueous Pt precursor than when Pt organic precursor was used.

Table 1. Nominal and experimental Pt/Co mole ratios; pH of BCME before and after $NaBH_4$ addition.

Sample	Pt/Co Mole Ratio		pH BCME	
	Nominal	Experimental (ICP-AES)	Before $NaBH_4$ Addition	After $NaBH_4$ Addition
PtAq-CoAq	0.30	1.55	1.57	6.36
PtAq-CoOi	0.12	0.43	6.60	8.66
PtOi-CoAq	0.76	0.17	4.10	10.17
PtOi-CoOi	0.30	0.12	7.34	10.30

In our synthesis, two scenarios should be considered: one in which both metallic precursors are present in the same nanophase and another in which they are individually dissolved in each liquid (water or isooctane). $NaBH_4$ is added to the BCME as a freshly prepared solution in water, and it is expected to incorporate to the aqueous nanochannels initially. It is well known that $NaBH_4$ hydrolysis is a pH-dependent reaction producing basic metaborate ions (which increase the solution pH) and hydrogen gas as products [29]. When $NaBH_4$ is in acidic media, H_2 is produced quickly. In contrast, at neutral or alkaline pH, $NaBH_4$ is much more stable, leading to slow H_2 release [30]. Therefore, the initial pH in the BCME greatly affects the presence of either the hydrolysis products or reducing BH_4^- ions as majoritarian species. On the other hand, the catalytic effect of several noble metal nanoparticles and their salts on the $NaBH_4$ hydrolysis reaction has been widely reported, which increases H_2 production [31–33].

On formation of H_2 gas, microbubbles that evolve in situ in the water phase may reach the oil phase due to the low surface tension inside the BCME conferred by the nonionic surfactant [34]. These microbubbles are surrounded by a surfactant monolayer. The hydrocarbon tails orient towards the inside H_2 bubble and EO chains towards the aqueous phase. It has been reported that EO chains are capable of preferentially solvate anions [35]; thus, we hypothesize that it is feasible that these H_2 microbubbles may be dragging some reducing species (BH_4^-) adsorbed on its surface, similarly to ion flotation phenomena [36], see Figure S2. Furthermore, H_2 itself is a slow reducing agent for platinum salts [37–39]. Thus, few nuclei of reduced metal in the oil phase are sufficient to act as seeds for superstructure growth when one of both reagents are present in this phase.

For the preparation of the PtAq-CoAq solid, the presence of $H_2PtCl_6.6H_2O$ as a water-soluble Pt precursor lowers the pH of the BCME prior to $NaBH_4$ addition (pH = 1.57). Under this condition, $NaBH_4$ hydrolysis is promoted; thus, H_2 and OH^- are produced, as evidenced by the increase of the pH after the recovery of the solids (i.e., 6.36). Both reducing species, as well as OH^- ions, become available in the same nanophase, resulting in a one-step reduction (Pt)/precipitation (Co_3O_4) inside the water nanochannels. In this case, the relatively high standard reduction potentials of the platinum precursor ($E^0_{[PtCl6]^{2-}/[PtCl4]^{2-}}$ = 0.68 V; $E^0_{[PtCl4]^{2-}/Pt}$ = 0.73 V) explain the higher Pt amount in the sample despite starting from a cobalt-rich mixture. In contrast, when using the Pt oil-soluble precursor (PtCOD) to obtain PtOi-CoAq, the microemulsion had an initial pH of 4.10, closer to BCME without any precursors. In this case, the presence of the reducing agent in the water nanochannels and the platinum precursor in the oil nanochannels suggests that the reduction reaction can only occur at the interface or by contact with the H_2 microbubbles dragging BH_4^- through the oil channels. Consequently, a lower production of metallic Pt is observed, as revealed in the lower Pt/Co ratio compared to the PtAq-CoAq solid.

On the other hand, the use of cobalt 2-ethylhexanoate as Co oil-precursor to obtain either PtAq-CoOi or PtOi-CoOi increases the initial pH up to neutral values due to the presence of 2-ethylhexanoate species at the interface [40]. For the PtAq-CoOi solid, the H^+ provided by the acidic platinum precursor neutralizes the unreacted hydroxyl species (produced from the $NaBH_4$ hydrolysis) causing a lower final pH in this system. It is worth noticing that the pH of the BCME barely changes after the synthesis of Pt-HS (initial pH = 0.57; final pH = 0.69). In the case of PtOi-CoOi, the experimental Pt/Co ratio is the lowest of all the samples, since platinum is in the oil phase and the BCME showed the highest initial pH value (Table 1); thus, the production of H_2 is lessened and the presence of reducing species passing through the oil phase is not favored.

Regarding the formation of Co species, metallic Co was not obtained. For all the compositions tested, the conditions were not appropriate for the reduction of Co^{2+} to metallic cobalt; instead, Co_3O_4 was formed. The formation of this oxide depends on the pH of the BCME; its formation is favored as the pH becomes more alkaline. As observed in Table 1, the final pH of the BCME for the PtAq-CoAq sample was 6.36, leading to a low Co_3O_4 content and the highest Pt/Co ratio obtained. The relatively low pH did not allow the formation of a large amount of Co_3O_4, leaving a considerable amount of precursor unreacted. In contrast, PtOi-CoOi, which started at pH = 7.34 and reached pH = 10.30 after $NaBH_4$ addition, allowed the formation of a larger amount of Co_3O_4. According to Carlo et al., 2015, Co particles were obtained with $NaBH_4$. It seems that in this work the simultaneous presence of Pt and Co precursors implied a competition, in which Pt^{4+} is reduced to Pt^0 and cobalt precipitates due to a consequent increase in pH as shown in Table 1. Actually, Co_3O_4 is the product obtained when NaOH or Oxalic acid are used [25]. Overall, we observed a significant dependence of initial and final pH values on the combination of precursors used.

Figure 3 shows the X-ray diffraction pattern of the PtAq-CoAq-HS compared to Pt-HS. Reference peaks for *fcc* Pt and cubic Co_3O_4 are included. Figure 3a shows the diffraction peaks located at 40.01°, 46.39° and 67.69°, these are assigned to the (111), (200) and (220) planes, respectively and were indexed to face-centered cubic (*fcc*) Pt (JCPDS file 98-002-1963). In Figure 3b, the broad diffraction peak at Bragg angle of 40.38° corresponds to Pt (111) and a possible contribution from Co_3O_4 (222) reflection, which is expected in PtAq-CoAq. Likewise, the diffraction peak at around 47.17° is ascribed to fcc crystalline Pt (200), and it closely coincides with cubic crystalline Co_3O_4 (400), while the peak at 68.39° is ascribed to fcc crystalline Pt (220), coincident with cubic crystalline Co_3O_4 (440). [Co_3O_4 JCPDS file number is 04-022-7366 (space group Fd-3m)]. The Pt (111) peak is broadened and shifted to a higher angle in comparison to the reference Pt and Pt-HS peaks. Following the Debye-Scherrer equation, the crystallite size was estimated for PtAq-CoAq to be 3.21 ± 0.45 nm, while the size of Pt-HS was estimated to be 4.81 ± 0.24 nm. The Pt and Co_3O_4 peaks superimposition, as well as the decrease in the crystallite size could be responsible for the broadening of the XRD pattern of PtAq-CoAq compared to Pt-HS. However, a possible microstrain in the Pt lattice should not be ruled out, due to replacement of some Pt by Co atoms, which have a lower atomic radius (r_{Co} = 1.52 Å vs. r_{Pt} = 1.77 Å) [41–44]. PtAq-CoAq exhibit a diffractogram coinciding to a greater extent with the reference Pt pattern, indicating that crystalline Pt is the main component in this sample, also in agreement with the ICP-AES results which revealed more Pt in the Pt-Co_3O_4-HSs synthesized with Pt aqueous soluble precursor. The diffractograms for the other Pt-Co_3O_4-HSs (PtAq-CoOi, PtOi-CoAq, and PtOi-CoOi) present broad, noisy, and undefined peaks (Figure S3).

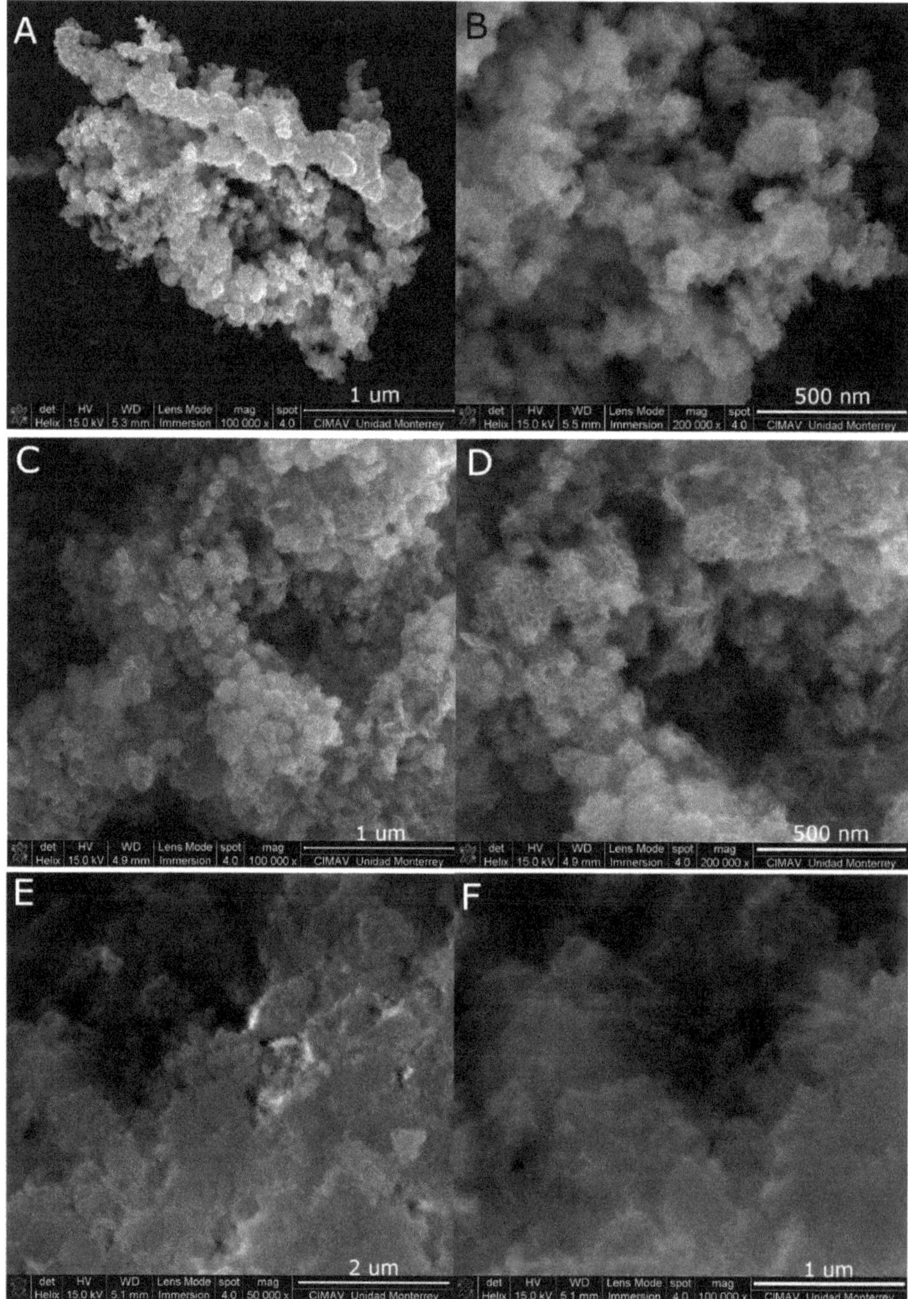

Figure 1. *Cont.*

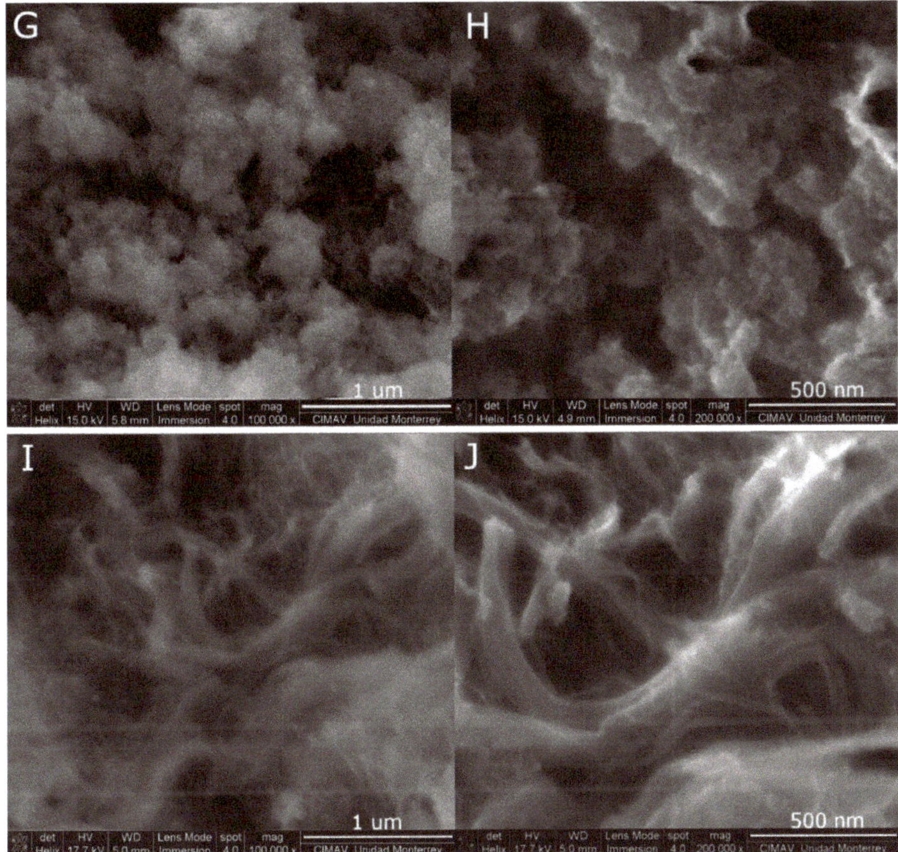

Figure 1. SEM images of Pt-HS (**A**,**B**), Pt-Co_3O_4(s)-HSs. PtAq-CoAq (**C**,**D**), PtAq-CoOi (**E**,**F**), PtOi-CoAq (**G**,**H**), PtOi-CoOi (**I**,**J**).

Since only one of the Pt-Co_3O_4-HSs presented a defined XRD pattern (PtAq-CoAq), these samples were analyzed by Raman Spectroscopy in order to confirm the nature of the cobalt phase. Figure 4 shows the Raman spectra of Pt-Co_3O_4-HSs; all samples presented the characteristic spectra for Co_3O_4 with a cubic spinel structure. Five Raman bands were observed at about 186, 463, 504, 601, and 662–668 cm^{-1} for all the samples, which could be ascribed to the $F_{2g}^{(1)}$, E_{2g}, $F_{2g}^{(2)}$, $F_{2g}^{(3)}$, and A_{1g} symmetry of crystalline cubic Co_3O_4 phase [45,46]. In particular, the Raman band at 186 cm^{-1} can be assigned to $F_{2g}^{(1)}$ symmetry in tetrahedral sites (CoO_4), which can be attributed to vibration of Co^{2+}-O^{2-}; while the band at 662–668 cm^{-1} can be assigned to A_{1g} species in octahedral sites (CoO_6), which can be attributed to vibration of Co^{3+}-O^{2-} [18].

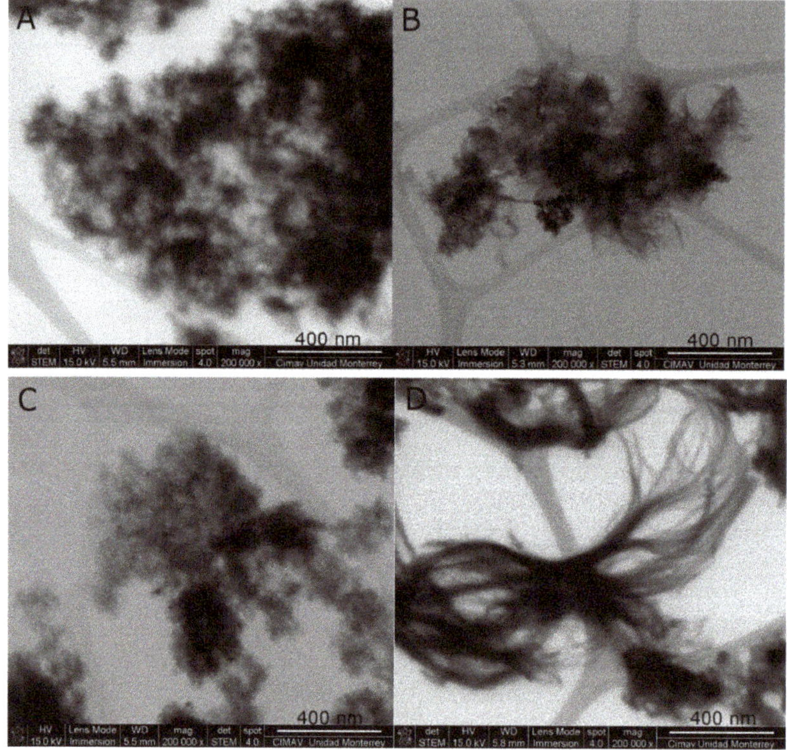

Figure 2. STEM images of Pt-Co_3O_4-HSs: PtAq-CoAq (**A**), PtAq-CoOi (**B**), PtOi-CoAq (**C**), PtOi-CoOi (**D**).

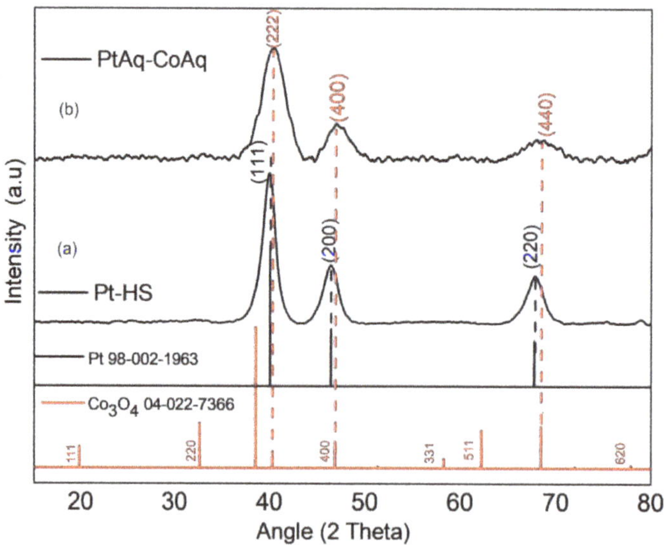

Figure 3. The XRD patterns obtain for Pt-HS (**a**) and PtAq-CoAq (**b**).

Furthermore, as observed in the inset of Figure 4, the position of the A_{1g} band varied consistently from 668 to 662 cm^{-1} depending on the sample. This shift towards lower frequencies has been associated with the increase in oxygen vacancies [46,47], although other studies attribute this shift to the amorphous state of the samples [48]. In our case, XRD spectra of PtAq-CoOi, PtOi-CoAq and PtOi-CoOi indicate that the samples are rather amorphous (Figure S3), although given the reducing conditions of the reaction, oxygen vacancies cannot be ruled out.

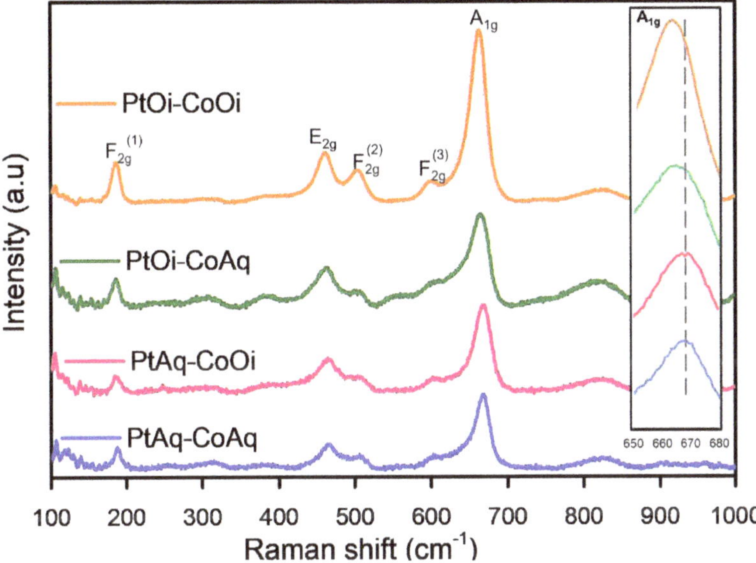

Figure 4. Raman spectra of Pt-Co_3O_4-HSs. The inset shows the shift of A_{1g} band towards lower frequency.

The surface composition and chemical oxidation states of Pt-Co_3O_4-HSs and Pt-HS was determined by X-ray photoelectron spectroscopy. Figure 5A shows the XPS survey spectra which revealed the expected photoelectron peaks of Pt, C, O and Co for Pt-Co_3O_4-HSs, while for Pt-HS, as expected, the peak due to Co was not found. Boron or chloride species from the precursors were also disregarded. The highest intensity for the Pt 4f photoelectron peak was found in Pt-HS, while the intensity of the O 1s peak in this sample was much lower as compared to the Pt-Co_3O_4-HSs solids.

The Pt 4f high-resolution spectra in Figure 5B show the corresponding spin–orbit coupling ($4f_{7/2}$ and $4f_{5/2}$) centered around 71 and 74 eV, respectively. For the Pt-HS, the deconvolution of the $4f_{7/2}$ peak showed the contribution of zerovalent platinum (Pt^0), as well as Pt^{2+} forming Pt-O and Pt-OH type-bonds (71.47, 72.39, and 73.17.4 eV) [49]. The binding energy for metallic Pt^0 (71.47 eV) is slightly higher than the value expected for bulk Pt (71.2 eV) [50], as well as the values obtained in the Pt-Co_3O_4-HS(s) samples. This is explained to be as a result of small cluster-size effect [16,51]. Regarding the platinum species existing in the Pt-Co_3O_4-HSs nanomaterials, the deconvoluted $4f_{7/2}$ peak revealed only two contributions assigned to Pt^0 and Pt^{2+}-O centered at 71.16, 72.09 eV for PtAq-CoAq; 71.06, 72.39 eV for PtAq-CoOi; 70.68, 72.31 eV for PtOi-CoAq and 71.00, 71.95 eV for PtOi-CoOi. As observed, the binding energy for Pt^0 species are lower in HS prepared using the oil soluble precursor.

On the other hand, the presence of oxidized platinum species, suggests that reduction is not fully completed under these conditions. In the case of the synthesis using chloroplatinic acid the reduction must be done from a Pt^{4+} oxidation state; while in the PtCOD precursor the Pt exists as

Pt^{2+}. The Pt^0/Pt^{2+} ratio was dependent on the combination of precursors used; for Pt-HS, which was prepared using aqueous Pt precursor $H_2PtCl_6 \bullet 6H_2O$, the Pt^0/Pt^{2+} ratio was 1.36; in comparison, for samples PtAq-CoAq and PtAq-CoOi, which were also prepared with the same precursor, the ratios were 0.96 and 0.7, respectively, indicating that the presence of Co precursor (and the formation of Co_3O_4) inhibited the reduction of Pt^{4+} to Pt^0. On the contrary, for PtOi-CoAq and PtOi-CoOi that were synthesized using the oil-soluble platinum precursor, the Pt^0/Pt^{2+} ratio is higher. In particular, the sample synthesized with both Pt and Co oil-soluble precursors, presented the highest Pt^0/Pt^{2+} ratio with a value of 8.12. According to these results, it seems easier to reduce the oil-soluble Pt precursor to Pt^0, despite the reducing species are less available in the oil phase. However, it should also be considered that in the case of synthesis using the platinum water-soluble precursor, the aqueous environment may favor stabilization of higher Pt oxidation state, in comparison with the oil environment.

The high-resolution spectra of Co 2p are shown in Figure 5C. The Co $2p_{3/2}$ and Co $2p_{1/2}$ doublet resulted from a contribution of four signals assigned to Co^{3+}, Co^{2+} species and satellite peaks [11,52]. The full peaks list with the corresponding assignation is shown in Table S1. The spin–orbit splitting of 15.82 eV for PtAq-CoAq; 15.75 eV for PtAq-CoOi; 15.56 eV for PtOi-CoAq; 15.78 eV for PtOi-CoOi are characteristic for Co_3O_4. The presence of CoO and Co_2O_3 can be disregarded as splitting values for these oxides are 16.00 and 15.00 eV, respectively [10]. Likewise, the binding energy at 778.00 eV, which indicates the presence of metallic Co^0, was not present in the XPS spectra of the synthesized nanomaterials [4,53]. Lower Co^{2+}/Co^{3+} ratio obtained in the case of the nanostructures prepared with both precursors dissolved in the same phase (Figure 5C) suggests that in these materials the oxidation of Co^{3+} to Co^{4+} through an electron-withdrawing inductive effect by Pt should be more favored.

In Figure 5D, each O 1s spectrum can be deconvoluted into three peaks. The peaks labeled as O1 and O2 with binding energies at 530.79 and 531.63 eV for PtAq-CoAq; 530.43 and 531.49 eV for PtAq-CoOi; 529.96 and 531.32 eV for PtOi-CoAq; 530.44 and 531.54 eV for PtOi-CoOi, are attributed to the lattice oxygen in Co-O of Co_3O_4 and defect sites with low oxygen coordination oxygen-vacancy, respectively. For Pt-HS, these peaks at 530 and 531.19 eV are attributed to the mentioned Pt-O and Pt-OH interaction. The peak labelled as O3 appears at 532.65 eV for PtAq-CoAq, 532.52 eV for PtAq-CoOi, 532.72 eV for Pt-HS is due to the oxygen atoms in -CO bonds, probably arising from the surface contamination with adventitious carbon. However, the contribution of -OH species should not be ruled out [54]. In the case of the peaks at 533.53 eV in PtOi-CoAq and 533.11 eV in PtOi-CoOi solids, it could be ascribed to adsorbed molecular water [6,19,55,56]. A higher O2:O1 ratio, as estimated from the ratio between the area of the peaks, may indicate a larger amount of surface oxygen vacancies [6,18,57,58]. The higher O2:O1 ratio were achieved in those samples prepared incorporating the precursors in the same phase: for PtOi-CoOi this value is 1.51 in agreement with Raman results (higher shift in the A_{1g} band), while for PtAq-CoAq this value is 1.41. This result also correlates with the lower Co^{2+}/Co^{3+} ratio found in these solids, and it is independent of the Co_3O_4 content, as the Pt/Co ratio is very different in both samples. The surface oxygen vacancies are useful in OER, and thus, these results would surely reflect in the different performance of the materials prepared. Therefore, having both precursors in the same phase competing for the reducing/precipitating species leads to higher surface defects in Co_3O_4 and a higher Co^{3+} population.

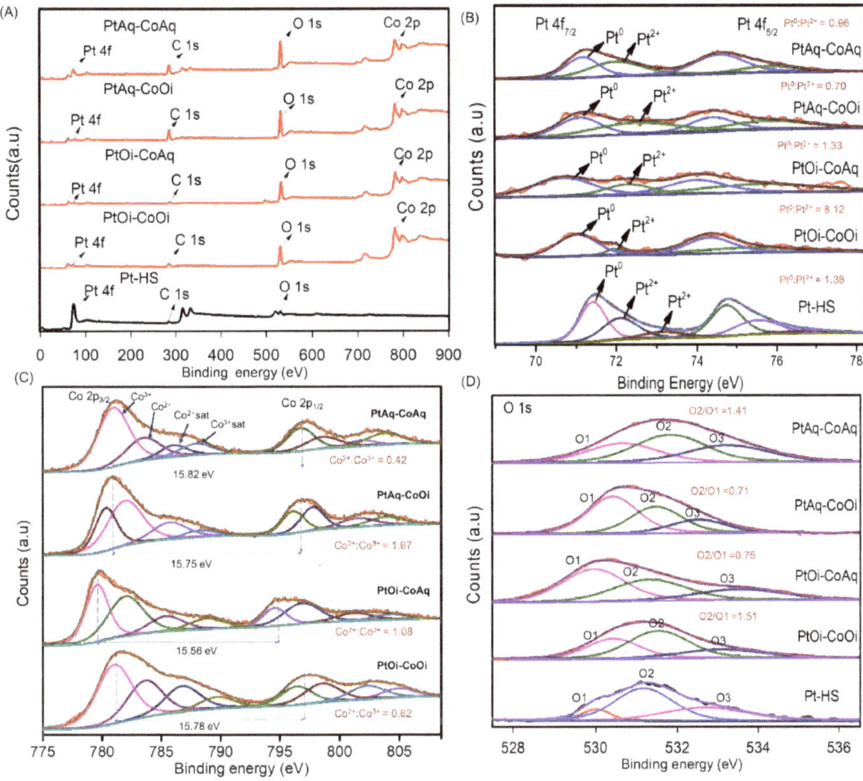

Figure 5. (**A**) XPS survey spectrum of Pt-Co_3O_4-HSs and Pt-HS, high-resolution spectra of (**B**) Pt 4f, (**C**) Co 2p, and (**D**) O 1s.

In this report, the water splitting capability of Pt-Co_3O_4-HSs and Pt-HS electrocatalysts was evaluated by investigating the OER activity through linear sweep voltammetry (LSV) in 0.1 M KOH with a sweep rate of 1 mV/s. The total catalysts loading of Pt-Co_3O_4-HSs and Pt-HS of 200 μg was maintained and drop-casted onto a glassy carbon electrode. As shown in Figure 6A, Pt-HS and Pt-Co_3O_4-HSs showed different OER performances. In Figure 6A inset, the LSV curve of the Pt-Co_3O_4-HSs shows anodic peaks at 1.15 V, 1.32 V, 1.19 V and 1.16 V vs. RHE for PtAq-CoAq, PtAq-CoOi, PtOi-CoAq and PtOi-CoOi, respectively, which correspond to Co^{2+} oxidation to Co^{3+} (Co^{2+}/Co^{3+}) [6,17,59,60]. This anodic peak precedes the onset of the OER and the sudden increase in the current density. However, Pt-HS did not show a distinct anodic peak for Pt oxidation, this process might be occurring in the same range at which OER happens on the surface of Pt-HS catalyst. The overpotential corresponding to a current density of 10 mA cm^{-2} (η_{10}) is of high relevance and a primary figure of merit for catalytic activity of oxygen evolution. It is a measure of the current density expected for an integrated solar water-splitting device operating at 10% solar-to-fuels efficiency. This is a common and widely used criterion for assessing OER activities [61,62]. Table 2 shows the overpotential corresponding to the current density at 10 mA cm^{-2} as well as the onset of oxygen evolution and Tafel slopes for Pt-Co_3O_4-HSs and Pt-HS. η_{10} for Pt-HS was calculated by extrapolation of the polarization curve.

Table 2. The overpotential at 10 mA cm^{-2}, onset of oxygen evolution and Tafel slopes for the electrocatalysts.

Electrocatalyst	Overpotential at 10 mA cm^{-2} (V) η_{10}	Onset (V) vs. RHE	Tafel Slope mV dec^{-1}
PtAq-CoAq	0.381	1.49	102
PtAq-CoOi	0.406	1.50	89
PtOi-CoAq	0.513	1.48	157
PtOi-CoOi	0.411	1.46	92
Pt-HS	0.627	1.56	117

The overpotential was calculated by subtracting water oxidation potential (1.229 V vs. RHE) from the applied potential [63]. In comparison, the Pt-Co_3O_4-HSs electrocatalysts synthesized in this work showed that PtAq-CoAq had η_{10} of 381 mV, while PtOi-CoAq had the highest η_{10} of 513 mV. PtAq-CoOi and PtOi-CoOi had overpotential value of 406 mV and 411 mV respectively. Pt-HS in comparison to the Pt-Co_3O_4-HSs, exhibited the highest overpotential of 627 mV to achieve a current density of 10 mA cm^{-2}. A good OER electrocatalyst is expected to produce a current density of 10 mA cm^{-2} at low overpotential. The high surface population of Co^{3+} species and the interconnected hierarchical structure in PtAq-CoAq could be favoring the Co^{3+} to Co^{4+} oxidation driven by the electron-withdrawing inductive effect of Pt, favoring the OER [14]. The potential at which oxygen evolution commences is called the onset potential and it varies between the electrocatalysts. According to the values in Table 2, PtOi-CoOi exhibited the lowest onset potential value of 1.46 V, while Pt-HS showed the onset at 1.56 V. On the other hand, the Tafel slope provides information associated with the rate determining steps and hence, it allows to compare the kinetics of the electrocatalytic reactions occurring on the surface of each material. The potential range at which the Tafel plot was calculated is labeled as the Tafel region in Figure 6A, while The Tafel slope of Pt-Co_3O_4-HSs and Pt-HS are shown in Figure 6B, these values are also listed in Table 2. PtAq-CoOi exhibited the lowest Tafel slope (89 mV dec^{-1}), which is synonymous with the catalyst having a faster kinetics to evolve oxygen from water as compared to Pt-HS which had a Tafel slope of 117 mV dec^{-1}. Tafel slopes near 120 mV dec^{-1} are observed when the surface species formed just before the rate-determining step are predominant; under alkaline conditions and assuming a single site mechanism it corresponds to the following equilibrium: $M + OH^- \leftrightarrow MOH + e^-$. However, when the surface adsorbed species produced in the early stage of the OER remains predominant the Tafel slope decreases [64]. Additionally, the presence and high ratio of Co^{2+}/Co^{3+} ions in PtAq-CoOi indicate an increase in the pre-oxidation of Co^{2+}, which enhances O_2-related electrocatalysis [18,56]. These factors could account for the lower Tafel slope of PtAq-CoOi as compared to Pt-HS.

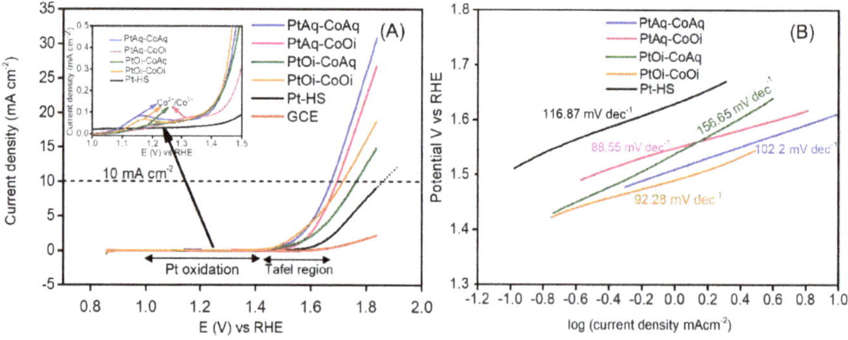

Figure 6. (**A**): LSV curves for OER activity of PtAq-CoAq, PtAq-CoOi, PtOi-CoAq, PtOi-CoOi and Pt-HS in 0.1 M KOH at a scan rate of 1 mV/s. (**B**) Tafel plots for PtAq-CoAq, PtAq-CoOi, PtOi-CoAq, PtOi-CoOi and Pt-HS.

However, in PtAq-CoAq the lowest Co^{2+}/Co^{3+} ratio was estimated which suggests a smaller amount of Co^{2+} (Figure 5C) is available to be effectively bounded by surface oxygen vacancies and oxidized to a higher valence [65]. This is suggested to be responsible for the higher Tafel slope of PtAq-CoAq as compared to PtAq-CoOi. Based on the morphological differences between interconnected needle-like PtAq-CoAq and nanoplatelet-like PtAq-CoOi, it is worthy to note that the electrocatalytic species in the nanoplatelet PtAq-CoOi (Figure 1E,F and Figure 2B) are sufficiently exposed to the electrolyte, thereby improving its surface reaction rate.

On the other hand, the atomic ratio of Co^{2+}/Co^{3+} and O2/O1 for PtOi-CoOi is 0.62 and 1.51, with Tafel slopes of 92 mV dec^{-1}; whereas PtOi-CoAq had a Co^{2+}/Co^{3+} ratio of 1.08 and O2/O1 of 0.75, with a Tafel slope of 157 mV dec^{-1}. In these two HSs, we identified that their morphology and arrangement (Figure 1G–J and Figure 2C,D), higher Co/Pt ratio (Table 1 ICP values), high surface oxygen vacancies (Raman) as well as atomic Co^{2+}/Co^{3+} ratio from XPS, had significant effect on the electrocatalytic activity. These favored the early OER onset potential, comparable overpotential for 10 mA cm^{-2} (Table 2) and lower Tafel slope for PtOi-CoOi when compared with PtOi-CoAq (Figure 6B). The fibrous morphological structure of PtOi-CoOi permits rapid electron transfer through the OER catalytic active sites [6,63].

From the LSV plots for the OER process, (Figure 6A,B) Pt-HS had the highest onset potential value, a Tafel slope higher than all the Pt-Co_3O_4(s) evaluated except PtOi-CoAq, whereas the potential corresponding to a current density of 10 mA cm^{-2} for Pt-HS was obtained by extrapolation. This shows that the catalytic performance of Pt-HS is significantly low without the presence of the mixed metal oxide. This highlights that the formation of interconnected structures in which Pt atom and mixed valence Co_3O_4 exist together enhances the performance of nanomaterials as electrocatalyst.

In general, the presence of Co_3O_4 in the nanomaterial using either oil soluble or aqueous soluble cobalt precursor showed an enhancement in the oxygen evolution reaction electrocatalytic activity of platinum. Therefore, for practical applications, the use of PtAq-CoAq is preferred not only because it has a unique interconnected needle-like morphology but because the overpotential to achieve the primary figure of merit η_{10} is low, earlier onset potential and a comparable Tafel slope (Table 3). In addition, the use of bicontinuous microemulsion as confined reaction media resulted in a good control of the morphology, as well as platinum and cobalt oxide incorporation into the nanomaterial, which are key for electrocatalytic applications.

The catalytic performance of the BCME synthesized Pt-Co_3O_4-HSs is comparable to reported literature for metal–metal oxides, and performed better than most metal oxides catalysts. This suggests that the Pt-Co_3O_4-HSs can serve as efficient electrocatalysts.

Chronoamperometry is a useful technique to evaluate the stability and durability of electrocatalysts [16]. Figure 7 shows the stability test of Pt-Co_3O_4-HSs and Pt-HS conducted at a potential of 1.50 V vs. RHE in 0.1 M KOH using chronoamperometry method for 4000 s. This potential was used because it is close to onset of OER as observed from the LSV (Figure 6A). The reaction is performed at constant potential with oxygen bubbles steadily generated on the surface of the electrocatalyst. The generated bubbles attach to the surface of the electrode, which resulted in decreased current density. The steady oxygen bubbles grow bigger with time and leave the catalyst surface. This increases current density immediately after the bubbles leave the active sites (Figure 7 inset). This process repeats throughout time under which chronoamperometry is performed. This gives a fluctuating current density-time curve. The decrease in the current density was 25%, 10.7%, 2.8%, 11.6% for PtAq-CoAq, PtAq-CoOi, PtOi-CoAq, PtOi-CoOi and 20.5% for Pt-HS after a long time of polarization. However, this change shows that the electrocatalyst are stable and can be used for long duration without substantial changes.

Table 3. Comparison of the Pt-Co_3O_4-HSs and Pt-HS electrocatalysts synthesized in this work versus reported literature.

Electrocatalysts	Electrolyte KOH	Onset Potential (V vs. RHE)	Overpotential (V) at 10 mA cm^{-2}	Tafel Slope (mV dec^{-1})	Reference
PtAq-CoAq	0.1 M	1.49	0.381	102	This work
PtAq-CoOi		1.50	0.406	89	This work
PtOi-CoAq		1.48	0.513	157	This work
PtOi-CoOi		1.46	0.411	92	This work
Pt-HS		1.56	0.627	117	This work
CoOx/Co-N-C Co-N-C	0.1 M	1.52 1.60	0.421 0.561	124 143	[56]
AuCuCo AuCu Pt/C	0.1 M	1.57 1.63 1.74	0.596 0.724 -	160 238 233	[63]
Co_3O_4/Co-NPC IrO_2	0.1 M	1.45 1.52	0.277 0.394	105 169	[24]
Pristine-Co_3O_4 Plasma-Co_3O_4	0.1 M	1.50 1.45	0.541 0.301	234 68	[11]
NixCoyO4/Co-NG NixCoyO4/Co-NG Pt	0.1 M	1.52 1.59 1.70	0.4 0.448 0.497	77 100 151	[15]
Pt-Co_3O_4 Pd-Co_3O_4 Au-Co_3O_4 Co_3O_4	0.1 M	1.383 1.388 1.395 1.481	0.254 - - -	58 61 68 72	[16]
$NiCo_2O_4$ $NiCo_2O_4$-G	0.1 M	1.593 1.523	0.424 0.564	205 161	[3]
$CaMnO_3$ H-Pt/$CaMnO_3$ Pt/$CaMnO_3$	0.1 M	1.66 1.58 1.58	 0.581 0.631	- - -	[66]
MnO_2/Ti CoP_3/Ti RuO_2/Ti MnO_2-CoP_3/Ti	1.0 M	1.6 1.54 1.45 1.545	0.408 0.322 0.24 0.288	241 276 - 65	[67]
CoTe Fe-CoTe	1.0 M	1.6 1.5	0.445 0.3	93 45	[68]
Au/C Co_3O_4/C Au-Co_3O_4/C	0.1 M	1.568 1.518 1.507	- - -	- - -	[10]
Fe_1Co_1-ONS Fe_1Co_1-ONP	0.1 M	1.46 1.56	0.308 0.4	37 72	[6]
Co_3O_4 Pd-Co_3O_4	0.1 M	1.481 1.388	- -	72 61	[4]
Co/N-C-800 Co/N-C-800-250 Co/N-C-800-450 Pt/C	0.1 M	1.53 1.53 1.52 1.65	0.371 0.424 0.398 0.629	61 116 74 170	[69]

NC: N-doped carbon, NPC: N-doped porous carbon matrix.

The mechanism of the OER using noble metal–cobalt oxide as electrocatalyst has been investigated [4,6,16,49]. It is reported that in metal–cobalt oxide catalysts, highly electronegative metals such as Pt facilitate the oxidation of Co^{3+} to produce Co^{4+} cations. This is majorly the rate limiting step, which highlights the importance of the noble metal. The electronegative metal allows for the easier formation of Co^{4+}, which is the active center for OER. Higher number of active centers on the electrode correspond to increase in the OER performance. Co^{4+} is more electrophilic, and will form intermediates by nucleophilic reactions with O species and OH^- at the reaction interface [16,70].

$$Co^{3+}{}_{(s)} \rightarrow Co^{4+}{}_{(s)} + e^- \quad (3)$$

$$Co^{4+}{}_{(s)} + OH^- \rightarrow Co^{4+}(OH)_{ad} + e^- \quad (4)$$

$$Co^{4+}(OH)_{ad} + OH^- \rightarrow Co^{4+}{}_{(s)} + 1/2O_2 + H_2O + e^- \quad (5)$$

The OER activity of Pt-Co_3O_4-HSs is better when compared to Pt-HS because multielectron transfer steps involved are promoted by the synergistic effect of the electrophilic Pt, the oxidation of Co^{3+} to Co^{4+} cation and surface oxygen vacancies. Additionally, the hierarchical structure of these materials made the active sites accessible for a higher electrocatalytic activity.

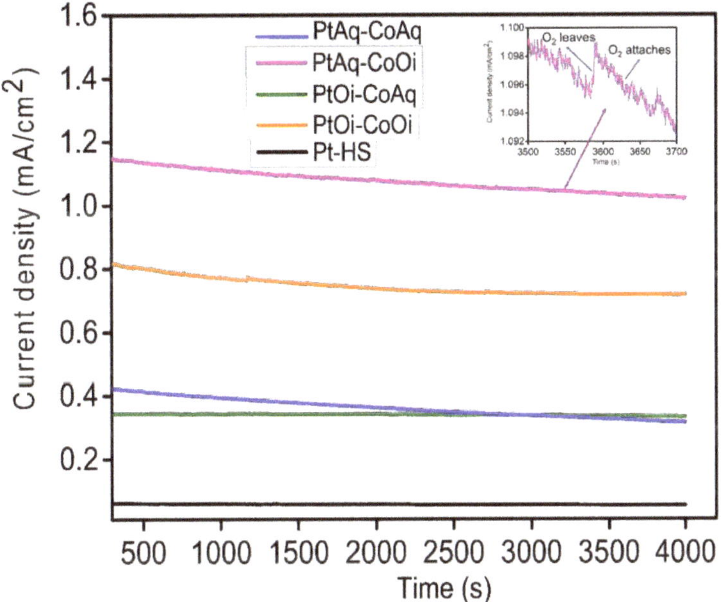

Figure 7. The current density-time curves for PtAq-CoAq, PtAq-CoOi, PtOi-CoAq, PtOi-CoOi and Pt-HS in 0.1 M KOH at 1.50 vs. RHE.

3. Materials and Methods

3.1. Chemicals

Cobalt (II) 2-ethylhexanoate (65 wt.% in mineral spirits, $CoC_{16}H_{30}O_4$, referred to as Co organic precursor), Cobalt (II) nitrate hexahydrate 99% ($Co(NO_3)_2.6H_2O$, referred to as Co aqueous precursor), (1,5-cyclooctadiene) dimethylplatinum (II) 97% ($C_{10}H_{16}Pt$, referred to as Pt organic precursor), Chloroplatinic acid hexahydrate 37.68% ($H_2PtCl_6.6H_2O$, referred to as Pt aqueous precursor), Isooctane 99% (C_8H_{18}), Sodium borohydride 99.99% ($NaBH_4$), KOH 90%, Nafion 5 wt.%, were purchased from Sigma Aldrich (Darmstadt, Germany). Synperonic^{TM} 91/5 ($C_{19}O_6H_{40}$) was purchased from CRODA. Isopropyl alcohol 99.9% (C_3H_8O) was supplied by Fisher Scientific (Hampton, NH, USA). Sulphuric acid 93–98% H_2SO_4 was purchased from Fermont.

3.2. Synthesis of Pt-Co_3O_4-HSs and Pt-HS

BCME was prepared using 52% of Isooctane, 27.36% of Synperonic^{TM} 91/5 and 20.64% of Milli-Q water. More details on this microemulsion system and its characterization can be found elsewhere [28]. Pt-Co_3O_4-HSs were produced by the reduction/precipitation of the aqueous or organic soluble precursors present in the BCME as follows:

I. Pt and Co aqueous precursors both present in the aqueous phase is referred as PtAq-CoAq
II. Pt and Co organic precursors both present in the organic phase is referred to as PtOi-CoOi
III. Pt aqueous precursor and Co organic precursor is referred to as PtAq-CoOi
IV. Pt organic precursor and Co aqueous precursor is referred to as PtOi-CoAq

The preparation of BCME was carried out in 20 mL glass vials. To prepare PtAq-CoAq, proper amounts of Pt and Co aqueous precursors were dissolved in Milli-Q water to give a final aqueous mixture containing 1 wt.% of Pt and 1 wt.% of Co. This was followed by the addition of Synperonic^{TM} 91/5 and Isooctane to obtain a final composition of 20.64% aqueous phase, 27.36% surfactant, and 52% oil phase. The mixture was vortexed for few seconds, resulting in a transparent, fluid, optically isotropic liquid phase and kept at 27 °C. A 100 µL of $NaBH_4$ solution with concentration equivalent to twice the total sum of Co and Pt moles, was prepared and added to the BCME within 3 s, and vortexed during 5 s. Then the reaction mixture was left to stand unstirred at 27 °C during 48 h prior to recovery and isolation of the nanomaterials. Detailed descriptions of the synthesis method for the other Pt-Co_3O_4-HSs are found in the Supplementary document. In brief, PtOi-CoOi, PtAq-CoOi and PtOi-CoAq are prepared by dissolving each precursor in the appropriate solvent, with the final mixture containing 1 wt.% of Pt and 1 wt.% of Co. The pH of BCMEs containing Pt and Co precursors before the addition of the $NaBH_4$ and after the recovery of the solids were measured using a Thermo Scientific Orion Star A211 pH Benchtop Meter.

3.3. Recovery and Isolation of Pt-Co_3O_4-HSs and Pt-HS

Isopropanol was added to the BCME containing the Pt-Co_3O_4-HSs in order to break the microemulsion and separate the solid; this was followed by washings with isopropanol, Milli-Q water and finally Isopropanol. For Pt-HS, a solution containing HNO_3: H_2O at 50:50 wt. ratio was used as additional washing solvent. Milli-Q water was then used to wash Pt-HS, until a neutral pH was recorded. The Pt-Co_3O_4-HSs and Pt-HS were then dried at room temperature, recovered and used for further characterizations.

3.4. Characterization of Pt-Co_3O_4-HSs and Pt-HS

The structure and morphologies of the synthesized Pt-Co_3O_4-HSs and Pt-HS were analyzed by scanning electron microscopy (SEM) and Scanning transmission electron microscopy (STEM) with a Field Emission Model 200 Nova NanoSEM at working voltages of 15–18 kV and equipped with an energy dispersive X-ray detector. The crystalline structure was characterized using a X-ray diffractometer PANalytical Empyrean equipped with a CuK$_\alpha$ cathode using a continuous scan mode from 10 to 80° (2θ) with 0.001 step size. The surface analysis was carried out using X-ray photoelectron spectroscopy (XPS) with a Thermo Fisher Scientific Escalab 250 Xi with an Al-K$_\alpha$ (1486.7 eV) X-ray source. Raman spectra were recorded with a Labram HR Evolution from Horiba, equipped with a 532 nm Nd: Yag laser (50 mW); the measurements were run with 10 s acquisition time, 50 accumulations. The metal content was measured using Inductively coupled plasma atomic emission spectroscopy (ICP-AES) with a Thermo Jarrell Ash iCAP 6000 equipment; appropriate sample treatment and dilution was carried out prior to ICP-AES measurement by open-beaker digestion with aqua regia.

3.5. Electrocatalytic Characteristics of Pt-Co_3O_4-HSs and Pt-HS

The electrocatalytic activity was evaluated using CH Instruments 440C Potentiostat/Galvanostat. The experiments were conducted at room temperature. A glassy carbon electrode (GCE) was used as the working electrode. It was polished using alumina powder suspensions of different particle sizes (1 μm, 0.3 μm and 0.1 μm). A three-electrode cell was assembled with a platinum mesh counter electrode and a Mercury/Mercurous oxide Reference Electrode (Hg|HgO, 20% KOH; 0.098 V vs. RHE). The oxygen evolution reaction was investigated by linear sweep voltammetry (LSV) in a range of 0 to 1.05 V vs. Hg|HgO in 0.1 M KOH at a scan rate of 1 mV/s.

The Pt-Co_3O_4-HSs and Pt-HS were dispersed in a 2:1 (Isopropanol: water) mixture, followed by a vortex mixing after ultrasonication. The concentration of the Pt-Co_3O_4-HSs and Pt-HS suspension prepared were 10 mg mL^{-1}. 20 μL of the solution containing the HSs was drop-casted onto the GCE electrode with an exposed cross section diameter of 7.9 mm. The HSs were allowed to dry at room temperature, after which 5 μL of a 0.05 wt.% Nafion solution was added to fix the HSs to the surface of the GCE. The stability of the HSs was investigated by chronoamperometry using a potential step of 650 mV vs. Hg|HgO in 0.1 M KOH for 4000 s. IR correction ($E_{corr} = E_{RHE} - iR_u$) was done on all the potentials, where E_{corr} is the iR_u-compensated potential, E_{RHE} is the potential obtained from experimentally measured potential and R_u is the resistance of the electrolyte determined by the iR compensation function of the CH Instruments 440 C electrochemical workstation.

4. Conclusions

The synthesis of Pt-Co_3O_4-HSs and Pt-HS using bicontinuous microemulsion was successfully carried out by a facile, fast, reproducible soft chemical method under ambient temperature and without further treatments. The characteristics, properties and performance of the obtained materials were dependent on the type of Pt and Co precursor (hydrophilic or lipophilic) and their combination; the best material was the one synthesized using both Pt and Co hydrophilic precursors. Oxygen evolution reaction performances of Pt-Co_3O_4-HSs were shown to be significantly faster and higher than on Pt-HS. The overpotential required to achieve 10 mA cm^{-2} when PtAq-CoAq catalyst was used was 381 mV, compared to 627 mV obtained when Pt-HS was used as catalyst. The onset potential of oxygen evolution on Pt-HS at 1.56 V vs. RHE was much higher than all the Pt-Co_3O_4-HSs (E_{onset} of PtAq-CoAq, PtAq-CoOi, PtOi-CoAq and PtOi-CoOi were 70, 60, 80, and 100 mV, respectively, more negative than that of Pt-HS). Additionally, the Tafel slopes of Pt-Co_3O_4-HSs are lower and the kinetics are faster than Pt-HS. The improvement in OER activity is due to the synergistic effect of electronegative Pt on Co_3O_4, which allows the formation of Co^{4+} species. This current research provides an opportunity to explore bicontinuous microemulsion as a structure-directing template for metal–metal oxide catalyst that are efficient in water electrolysis.

Supplementary Materials: The following are available online at http://www.mdpi.com/2073-4344/10/11/1311/s1, Figure S1: EDS images of Pt-Co3O4(s): PtAq-CoAq (A), PtAq-CoOi (B), PtOi-CoAq (C), PtOi-CoOi (D), Pt HS, Figure S2: Schematic representation of the H_2 microbubbles evolved in situ at the water channels, surrounded by surfactant molecules and dragging reducing species to the oil channels, Figure S3: The XRD pattern of Pt-Co3O4-HSs: PtOi-CoOi (a), PtAq-CoOi (b), PtOi-CoAq (c), Table S1: Table showing the peaks derive from the deconvolution of the Co 2p3/2 peak. The signals were assigned to Co^{3+} and Co^{2+} species with the corresponding satellite signals (shake-up).

Author Contributions: E.T.A.: methodology, validation, formal analysis, investigation and writing original draft, writing-review & editing; E.G.-V.: methodology and investigation; K.M.F.: methodology, investigation, writing-review & editing; M.V.: conceptualization, resources, writing-review & editing, supervision; M.S.-D.: conceptualization, resources, writing-review & editing, supervision, project administration and funding acquisition. All authors have read and agreed to the published version of the manuscript.

Funding: This research and the APC were funded by CONACYT (Fronteras de la Ciencia) grant number FC 2016/1700.

Acknowledgments: E.T. Adesuji and K.M. Fuentes acknowledge financial support from CONACYT (Ph.D. CONACYT Grant and postdoctoral grant from FC 2016/1700, respectively). M. Videa thanks the School of Engineering and Sciences at Tecnológico de Monterrey for partial funding provided through the Research Chair of Photonics and Quantum Systems. The authors acknowledge Francisco E. Longoria, J. Alejandro Arizpe, Nayely Pineda and L. Gerardo Silva (CIMAV Monterrey) for their help with XRD, HRTEM-STEM and Raman, SEM, and XPS measurements, respectively. We also thank Victor A. Luna Flores and Miguel A. Esneider Alcala for their help in cutting the electrodes.

Conflicts of Interest: The authors declare no conflict of interest.

References

1. Ozoemena, K.I. Nanostructured platinum-free electrocatalysts in alkaline direct alcohol fuel cells: Catalyst design, principles and applications. *RSC Adv.* **2016**, *6*, 89523–89550. [CrossRef]
2. Lim, T.; Sung, M.; Kim, J. Oxygen evolution reaction at microporous pt layers: Differentiated electrochemical activity between acidic and basic media. *Sci. Rep.* **2017**, *7*, 15382. [CrossRef] [PubMed]
3. Lee, D.U.; Kim, B.J.; Chen, Z. One-pot synthesis of a mesoporous nico2o4 nanoplatelet and graphene hybrid and its oxygen reduction and evolution activities as an efficient bi-functional electrocatalyst. *J. Mater. Chem. A* **2013**, *1*, 4754–4762. [CrossRef]
4. Qu, Q.; Zhang, J.H.; Wang, J.; Li, Q.Y.; Xu, C.W.; Lu, X. Three-dimensional ordered mesoporous co3o4 enhanced by pd for oxygen evolution reaction. *Sci. Rep.* **2017**, *7*, 41542. [CrossRef] [PubMed]
5. Gao, M.; Sheng, W.; Zhuang, Z.; Fang, Q.; Gu, S.; Jiang, J.; Yan, Y. Efficient water oxidation using nanostructured alpha-nickel-hydroxide as an electrocatalyst. *J. Am. Chem. Soc.* **2014**, *136*, 7077–7084. [CrossRef] [PubMed]
6. Zhuang, L.; Ge, L.; Yang, Y.; Li, M.; Jia, Y.; Yao, X.; Zhu, Z. Ultrathin iron-cobalt oxide nanosheets with abundant oxygen vacancies for the oxygen evolution reaction. *Adv. Mater.* **2017**, *29*. [CrossRef] [PubMed]
7. Bezerra, C.W.; Zhang, L.; Lee, K.; Liu, H.; Marques, A.L.; Marques, E.P.; Wang, H.; Zhang, J. A review of fe–n/c and co–n/c catalysts for the oxygen reduction reaction. *Electrochim. Acta* **2008**, *53*, 4937–4951. [CrossRef]
8. Wu, L.; Liu, Z.; Xu, M.; Zhang, J.; Yang, X.; Huang, Y.; Lin, J.; Sun, D.; Xu, L.; Tang, Y. Facile synthesis of ultrathin pd–pt alloy nanowires as highly active and durable catalysts for oxygen reduction reaction. *Int. J. Hydrogen Energy* **2016**, *41*, 6805–6813. [CrossRef]
9. Yuan, N.; Jiang, Q.; Li, J.; Tang, J. A review on non-noble metal based electrocatalysis for the oxygen evolution reaction. *Arab. J. Chem.* **2020**, *13*, 4294–4309. [CrossRef]
10. Li, Z.-Y.; Ye, K.-H.; Zhong, Q.-S.; Zhang, C.-J.; Shi, S.-T.; Xu, C.-W. Au-co3o4/c as an efficient electrocatalyst for the oxygen evolution reaction. *ChemPlusChem* **2014**, *79*, 1569–1572. [CrossRef]
11. Xu, L.; Jiang, Q.; Xiao, Z.; Li, X.; Huo, J.; Wang, S.; Dai, L. Plasma-engraved co3 o4 nanosheets with oxygen vacancies and high surface area for the oxygen evolution reaction. *Angew. Chem. Int. Ed. Engl.* **2016**, *55*, 5277–5281. [CrossRef]
12. Ceyssens, F.; Sree, S.P.; Martens, J.; Puers, R. Fabrication of nanostructured platinum with multilevel porosity for low impedance biomedical recording and stimulation electrodes. *Procedia Eng.* **2015**, *120*, 355–359. [CrossRef]
13. McAlpin, J.G.; Surendranath, Y.; Dinca, M.; Stich, T.A.; Stoian, S.A.; Casey, W.H.; Nocera, D.G.; Britt, R.D. Epr evidence for co (iv) species produced during water oxidation at neutral ph. *J. Am. Chem. Soc.* **2010**, *132*, 6882–6883. [CrossRef]
14. Yeo, B.S.; Bell, A.T. Enhanced activity of gold-supported cobalt oxide for the electrochemical evolution of oxygen. *J. Am. Chem. Soc.* **2011**, *133*, 5587–5593. [CrossRef]
15. Hao, Y.; Xu, Y.; Liu, J.; Sun, X. Nickel–cobalt oxides supported on co/n decorated graphene as an excellent bifunctional oxygen catalyst. *J. Mater. Chem. A* **2017**, *5*, 5594–5600. [CrossRef]
16. Qu, Q.; Pan, G.-L.; Lin, Y.-T.; Xu, C.-W. Boosting the electrocatalytic performance of pt, pd and au embedded within mesoporous cobalt oxide for oxygen evolution reaction. *Int. J. Hydrogen Energy* **2018**, *43*, 14252–14264. [CrossRef]
17. Zhuang, Z.; Sheng, W.; Yan, Y. Synthesis of monodispere au@co3o4 core-shell nanocrystals and their enhanced catalytic activity for oxygen evolution reaction. *Adv. Mater.* **2014**, *26*, 3950–3955. [CrossRef]
18. Xiao, Z.; Huang, Y.-C.; Dong, C.-L.; Xie, C.; Liu, Z.; Du, S.; Chen, W.; Yan, D.; Tao, L.; Shu, Z.; et al. Operando identification of the dynamic behavior of oxygen vacancy-rich co3o4 for oxygen evolution reaction. *J. Am. Chem. Soc.* **2020**, *142*, 12087–12095. [CrossRef]

19. Chen, Y.; Hu, L.; Min, Y.; Zhang, Y. Facile synthesis of noble metal (au,pt)/co3o4 nanocomposite microspheres and their catalytic activity. *Mater. Chem. Phys.* **2010**, *124*, 1166–1171. [CrossRef]
20. Grewe, T.; Deng, X.; Tüysüz, H. Influence of fe doping on structure and water oxidation activity of nanocast co3o4. *Chem. Mater.* **2014**, *26*, 3162–3168. [CrossRef]
21. Lu, X.; Ng, Y.H.; Zhao, C. Gold nanoparticles embedded within mesoporous cobalt oxide enhance electrochemical oxygen evolution. *ChemSusChem* **2014**, *7*, 82–86. [CrossRef]
22. Rosen, J.; Hutchings, G.S.; Jiao, F. Ordered mesoporous cobalt oxide as highly efficient oxygen evolution catalyst. *J. Am. Chem. Soc.* **2013**, *135*, 4516–4521. [CrossRef]
23. Deng, X.; Ozturk, S.; Weidenthaler, C.; Tuysuz, H. Iron-induced activation of ordered mesoporous nickel cobalt oxide electrocatalyst for the oxygen evolution reaction. *ACS Appl. Mater. Interfaces* **2017**, *9*, 21225–21233. [CrossRef]
24. Zhan, T.; Lu, S.; Liu, X.; Teng, H.; Hou, W. Alginate derived co3o4/co nanoparticles decorated in n-doped porous carbon as an efficient bifunctional catalyst for oxygen evolution and reduction reactions. *Electrochim. Acta* **2018**, *265*, 681–689. [CrossRef]
25. Carlo, G.D.; Lualdi, M.; Venezia, A.M.; Boutonnet, M.; Sanchez-Dominguez, M. Design of cobalt nanoparticles with tailored structural and morphological properties via o/w and w/o microemulsions and their deposition onto silica. *Catalysts* **2015**, *5*, 442–459. [CrossRef]
26. König, R.Y.; Schwarze, M.; Schomäcker, R.; Stubenrauch, C. Catalytic activity of mono-and bi-metallic nanoparticles synthesized via microemulsions. *Catalysts* **2014**, *4*, 256–275. [CrossRef]
27. Serrà, A.; Vallés, E. Microemulsion-based one-step electrochemical fabrication of mesoporous catalysts. *Catalysts* **2018**, *8*, 395. [CrossRef]
28. Adesuji, E.T.; Khalil-Cruz, L.E.; Videa, M.; Sánchez-Domínguez, M. From nano to macro: Hierarchical platinum superstructures synthesized using bicontinuous microemulsion for hydrogen evolution reaction. *Electrochim. Acta* **2020**, *354*, 136608. [CrossRef]
29. Schlesinger, H.I.; Brown, H.C.; Finholt, A.E.; Gilbreath, J.R.; Hoekstra, H.R.; Hyde, E.K. Sodium borohydride, its hydrolysis and its use as a reducing agent and in the generation of hydrogen1. *J. Am. Chem. Soc.* **1953**, *75*, 215–219. [CrossRef]
30. Liu, B.H.; Li, Z.P. A review: Hydrogen generation from borohydride hydrolysis reaction. *J. Power Sources* **2009**, *187*, 527–534. [CrossRef]
31. Brack, P.; Dann, S.E.; Wijayantha, K.G.U. Heterogeneous and homogenous catalysts for hydrogen generation by hydrolysis of aqueous sodium borohydride (nabh4) solutions. *Energy Sci. Eng.* **2015**, *3*, 174–188. [CrossRef]
32. Demirci, U.B.; Miele, P. Reaction mechanisms of the hydrolysis of sodium borohydride: A discussion focusing on cobalt-based catalysts. *Comptes Rendus Chim.* **2014**, *17*, 707–716. [CrossRef]
33. Kaufman, C.M.; Sen, B. Hydrogen generation by hydrolysis of sodium tetrahydroborate: Effects of acids and transition metals and their salts. *J. Chem. Soc. Dalton Trans.* **1985**, 307–313. [CrossRef]
34. Folgueiras-Amador, A.A.; Jolley, K.E.; Birkin, P.R.; Brown, R.C.D.; Pletcher, D.; Pickering, S.; Sharabi, M.; de Frutos, O.; Mateos, C.; Rincón, J.A. The influence of non-ionic surfactants on electrosynthesis in extended channel, narrow gap electrolysis cells. *Electrochem. Commun.* **2019**, *100*, 6–10. [CrossRef]
35. Ohki, T.; Harada, M.; Okada, T. Solvation of ions in hydrophilic layer of polyoxyethylated nonionic micelle. Cooperative approach by electrophoresis and ion-transfer voltammetry. *J. Phys. Chem. B* **2006**, *110*, 15486–15492. [CrossRef]
36. Chang, L.; Cao, Y.; Fan, G.; Li, C.; Peng, W. A review of the applications of ion floatation: Wastewater treatment, mineral beneficiation and hydrometallurgy. *RSC Adv.* **2019**, *9*, 20226–20239. [CrossRef]
37. Boutonnet, M.; Kizling, J.; Stenius, P.; Maire, G. The preparation of monodisperse colloidal metal particles from microemulsions. *Colloids Surf.* **1982**, *5*, 209–225. [CrossRef]
38. Henglein, A.; Giersig, M. Reduction of pt(ii) by h2: Effects of citrate and naoh and reaction mechanism. *J. Phys. Chem. B* **2000**, *104*, 6767–6772. [CrossRef]
39. Zeng, J. A simple eco-friendly solution phase reduction method for the synthesis of polyhedra platinum nanoparticles with high catalytic activity for methanol electrooxidation. *J. Mater. Chem.* **2012**, *22*, 3170–3176. [CrossRef]
40. Sanchez-Dominguez, M.; Pemartin, K.; Boutonnet, M. Preparation of inorganic nanoparticles in oil-in-water microemulsions: A soft and versatile approach. *Curr. Opin. Colloid Interface Sci.* **2012**, *17*, 297–305. [CrossRef]

41. Huang, H.; Hu, X.; Zhang, J.; Su, N.; Cheng, J. Facile fabrication of platinum-cobalt alloy nanoparticles with enhanced electrocatalytic activity for a methanol oxidation reaction. *Sci. Rep.* **2017**, *7*, 45555. [CrossRef]
42. Chattot, R.; Asset, T.; Bordet, P.; Drnec, J.; Dubau, L.; Maillard, F. Beyond strain and ligand effects: Microstrain-induced enhancement of the oxygen reduction reaction kinetics on various ptni/c nanostructures. *ACS Catal.* **2017**, *7*, 398–408. [CrossRef]
43. Maniammal, K.; Madhu, G.; Biju, V. X-ray diffraction line profile analysis of nanostructured nickel oxide: Shape factor and convolution of crystallite size and microstrain contributions. *Phys. E Low Dimens. Syst. Nanostruct.* **2017**, *85*, 214–222. [CrossRef]
44. Clementi, E.; Raimondi, D.L.; Reinhardt, W.P. Atomic screening constants from scf functions. Ii. Atoms with 37 to 86 electrons. *J. Chem. Phys.* **1967**, *47*, 1300–1307. [CrossRef]
45. Wang, J.; Gao, R.; Zhou, D.; Chen, Z.; Wu, Z.; Schumacher, G.; Hu, Z.; Liu, X. Boosting the electrocatalytic activity of co3o4 nanosheets for a li-o2 battery through modulating inner oxygen vacancy and exterior CO3+/CO2+ ratio. *ACS Catal.* **2017**, *7*, 6533–6541. [CrossRef]
46. Wang, X.; Li, X.; Mu, J.; Fan, S.; Chen, X.; Wang, L.; Yin, Z.; Tadé, M.; Liu, S. Oxygen vacancy-rich porous co3o4 nanosheets toward boosted no reduction by co and co oxidation: Insights into the structure–activity relationship and performance enhancement mechanism. *ACS Appl. Mater. Interfaces* **2019**, *11*, 41988–41999. [CrossRef] [PubMed]
47. Zhao, Q.; Fu, L.; Jiang, D.; Ouyang, J.; Hu, Y.; Yang, H.; Xi, Y. Nanoclay-modulated oxygen vacancies of metal oxide. *Commun. Chem.* **2019**, *2*, 11. [CrossRef]
48. Khajehbashi, S.M.B.; Li, J.; Wang, M.; Xu, L.; Zhao, K.; Wei, Q.; Shi, C.; Tang, C.; Huang, L.; Wang, Z.; et al. A crystalline/amorphous cobalt(ii,iii) oxide hybrid electrocatalyst for lithium-air batteries. *Energy Technol.* **2017**, *5*, 568–579. [CrossRef]
49. Li, Z.-Y.; Shi, S.-T.; Zhong, Q.-S.; Zhang, C.-J.; Xu, C.-W. Pt-mn 3 o 4/c as efficient electrocatalyst for oxygen evolution reaction in water electrolysis. *Electrochim. Acta* **2014**, *146*, 119–124. [CrossRef]
50. Peuckert, M.; Bonzel, H.P. Characterization of oxidized platinum surfaces by X-ray photoelectron spectroscopy. *Surf. Sci.* **1984**, *145*, 239–259. [CrossRef]
51. Eberhardt, W.; Fayet, P.; Cox, D.M.; Fu, Z.; Kaldor, A.; Sherwood, R.; Sondericker, D. Photoemission from mass-selected monodispersed pt clusters. *Phys. Rev. Lett.* **1990**, *64*, 780–783. [CrossRef]
52. Alexander, V.N.; Anna, K.; Stephen, W.G.; Cedric, J.P. Nist X-ray photoelectron spectroscopy database. In *NIST Standard Reference Database 20, Version 4.1*; National Institute of Standards and Technology: Gaithersburg, MD, USA, 2012.
53. Li, Z.-Y.; Liu, Z.-L.; Liang, J.-C.; Xu, C.-W.; Lu, X. Facile synthesis of pd–mn3o4/c as high-efficient electrocatalyst for oxygen evolution reaction. *J. Mater. Chem. A* **2014**, *2*, 18236–18240. [CrossRef]
54. Payne, B.P.; Biesinger, M.C.; McIntyre, N.S. X-ray photoelectron spectroscopy studies of reactions on chromium metal and chromium oxide surfaces. *J. Electron. Spectrosc. Relat. Phenom.* **2011**, *184*, 29–37. [CrossRef]
55. Aricò, A.S.; Shukla, A.K.; Kim, H.; Park, S.; Min, M.; Antonucci, V. An xps study on oxidation states of pt and its alloys with co and cr and its relevance to electroreduction of oxygen. *Appl. Surf. Sci.* **2001**, *172*, 33–40. [CrossRef]
56. Xin, W.-L.; Lu, K.-K.; Zhu, D.-R.; Zeng, H.-B.; Zhang, X.-J.; Marks, R.-S.; Shan, D. Highly reactive n,n′-carbonyldiimidazole-tailored bifunctional electrocatalyst for oxygen reduction and oxygen evolution. *Electrochim. Acta* **2019**, *307*, 375–384. [CrossRef]
57. Kuznetsov, D.A.; Naeem, M.A.; Kumar, P.V.; Abdala, P.M.; Fedorov, A.; Müller, C.R. Tailoring lattice oxygen binding in ruthenium pyrochlores to enhance oxygen evolution activity. *J. Am. Chem. Soc.* **2020**, *142*, 7883–7888. [CrossRef]
58. Ramos-Moore, E.; Ferrari, P.; Diaz-Droguett, D.E.; Lederman, D.; Evans, J.T. Raman and X-ray photoelectron spectroscopy study of ferroelectric switching in pb(nb,zr,ti)o3 thin films. *J. Appl. Phys.* **2012**, *111*, 014108. [CrossRef]
59. Bajdich, M.; Garcia-Mota, M.; Vojvodic, A.; Norskov, J.K.; Bell, A.T. Theoretical investigation of the activity of cobalt oxides for the electrochemical oxidation of water. *J. Am. Chem. Soc.* **2013**, *135*, 13521–13530. [CrossRef]
60. Wang, Y.; Zhou, T.; Jiang, K.; Da, P.; Peng, Z.; Tang, J.; Kong, B.; Cai, W.-B.; Yang, Z.; Zheng, G. Reduced mesoporous co3o4nanowires as efficient water oxidation electrocatalysts and supercapacitor electrodes. *Adv. Energy Mater.* **2014**, *4*. [CrossRef]

61. Kunhiraman, A.K. Hydrogen evolution reaction catalyzed by platinum nanoislands decorated on three-dimensional nanocarbon hybrid. *Ionics* **2019**, *25*, 3787–3797. [CrossRef]
62. McCrory, C.C.L.; Jung, S.; Ferrer, I.M.; Chatman, S.M.; Peters, J.C.; Jaramillo, T.F. Benchmarking hydrogen evolving reaction and oxygen evolving reaction electrocatalysts for solar water splitting devices. *J. Am. Chem. Soc.* **2015**, *137*, 4347–4357. [CrossRef]
63. Gong, H.; Zhang, W.; Li, F.; Yang, R. Enhanced electrocatalytic performance of self-supported aucuco for oxygen reduction and evolution reactions. *Electrochim. Acta* **2017**, *252*, 261–267. [CrossRef]
64. Shinagawa, T.; Garcia-Esparza, A.T.; Takanabe, K. Insight on tafel slopes from a microkinetic analysis of aqueous electrocatalysis for energy conversion. *Sci. Rep.* **2015**, *5*, 13801. [CrossRef]
65. Wang, Z.; Wang, W.; Zhang, L.; Jiang, D. Surface oxygen vacancies on co3o4 mediated catalytic formaldehyde oxidation at room temperature. *Catal. Sci. Technol.* **2016**, *6*, 3845–3853. [CrossRef]
66. Han, X.; Cheng, F.; Zhang, T.; Yang, J.; Hu, Y.; Chen, J. Hydrogenated uniform pt clusters supported on porous camno(3) as a bifunctional electrocatalyst for enhanced oxygen reduction and evolution. *Adv. Mater.* **2014**, *26*, 2047–2051. [CrossRef] [PubMed]
67. Xiong, X.; Ji, Y.; Xie, M.; You, C.; Yang, L.; Liu, Z.; Asiri, A.M.; Sun, X. Mno2-cop3 nanowires array: An efficient electrocatalyst for alkaline oxygen evolution reaction with enhanced activity. *Electrochem. Commun.* **2018**, *86*, 161–165. [CrossRef]
68. Zhong, L.; Bao, Y.; Feng, L. Fe-doping effect on cote catalyst with greatly boosted intrinsic activity for electrochemical oxygen evolution reaction. *Electrochim. Acta* **2019**, 321. [CrossRef]
69. Su, Y.; Zhu, Y.; Jiang, H.; Shen, J.; Yang, X.; Zou, W.; Chen, J.; Li, C. Cobalt nanoparticles embedded in n-doped carbon as an efficient bifunctional electrocatalyst for oxygen reduction and evolution reactions. *Nanoscale* **2014**, *6*, 15080–15089. [CrossRef] [PubMed]
70. Zhu, H.; Jiang, R.; Chen, X.; Chen, Y.; Wang, L. 3d nickel-cobalt diselenide nanonetwork for highly efficient oxygen evolution. *Sci. Bull.* **2017**, *62*, 1373–1379. [CrossRef]

Publisher's Note: MDPI stays neutral with regard to jurisdictional claims in published maps and institutional affiliations.

© 2020 by the authors. Licensee MDPI, Basel, Switzerland. This article is an open access article distributed under the terms and conditions of the Creative Commons Attribution (CC BY) license (http://creativecommons.org/licenses/by/4.0/).

Article

Substitution of Co with Ni in Co/Al_2O_3 Catalysts for Fischer–Tropsch Synthesis

Michela Martinelli [1], Sai Charan Karuturi [2], Richard Garcia [3], Caleb D. Watson [3], Wilson D. Shafer [4], Donald C. Cronauer [5], A. Jeremy Kropf [5], Christopher L. Marshall [5] and Gary Jacobs [1,3,*]

1 University of Kentucky Center for Applied Energy Research, 2540 Research Park Drive, Lexington, KY 40511, USA; michela.martinelli@uky.edu
2 Department of Mechanical Engineering, University of Texas at San Antonio, One UTSA Circle, San Antonio, TX 78249, USA; saicharan.karuturi@my.utsa.edu
3 Department of Biomedical Engineering and Chemical Engineering, University of Texas at San Antonio, One UTSA Circle, San Antonio, TX 78249, USA; g123richard@gmail.com (R.G.); caleb.watson378@gmail.com (C.D.W.)
4 Department of Chemistry, Asbury University, One Macklem Drive, Wilmore, KY 40390, USA; Wilson.Shafer@asbury.edu
5 Argonne National Laboratory, Argonne, IL 60439, USA; dccronauer@anl.gov (D.C.C.); kropf@anl.gov (A.J.K.); marshall@anl.gov (C.L.M.)
* Correspondence: gary.jacobs@utsa.edu; Tel.: +1-210-458-7080

Received: 22 February 2020; Accepted: 15 March 2020; Published: 17 March 2020 

Abstract: The effect of cobalt substitution with nickel was investigated for the Fischer–Tropsch synthesis reaction. Catalysts having different Ni/Co ratios were prepared by aqueous incipient wetness co-impregnation, characterized, and tested using a continuously stirred tank reactor (CSTR) for more than 200 h. The addition of nickel did not significantly modify the morphological properties measured. XRD, STEM, and TPR-XANES results showed intimate contact between nickel and cobalt, strongly suggesting the formation of a Co-Ni solid oxide solution in each case. Moreover, TPR-XANES indicated that nickel addition improves the cobalt reducibility. This may be due to H_2 dissociation and spillover, but is more likely the results of a chemical effect of intimate contact between Co and Ni resulting in Co-Ni alloying after activation. FTS testing revealed a lower initial activity when nickel was added. However, CO conversion continuously increased with time on-stream until a steady-state value (34%–37% depending on Ni/Co ratio) was achieved, which was very close to the value observed for undoped Co/Al_2O_3. This trend suggests nickel can stabilize cobalt nanoparticles even at a lower weight percentage of Co. Currently, the cobalt price is 2.13 times the price of nickel. Thus, comparing the activity/price, the catalyst with a Ni/Co ratio of 25/75 has better performance than the unpromoted catalyst. Finally, nickel-promoted catalysts exhibited slightly higher initial selectivity for light hydrocarbons, but this difference typically diminished with time on-stream; once leveling off in conversion was achieved, the C_5+ selectivities were similar (≈ 80%) for Ni/Co ratios up to 10/90, and only slightly lower (≈ 77%) at Ni/Co of 25/75.

Keywords: Fischer–Tropsch synthesis; bimetallic catalyst; cobalt-nickel alloys; TPR-XANES/EXAFS

1. Introduction

Fischer–Tropsch synthesis (FTS) is a catalytic reaction which converts syngas, a mixture of CO and H_2 derived from natural gas, coal, and/or biomass, to high quality fuels. The active metals for FTS are iron, cobalt, nickel, and ruthenium. Among these metals, ruthenium is the most active. However, its application for large-scale FTS plants is impractical because of low abundance and very high cost [1]. In contrast, nickel is cheaper, but high selectivities for short-chained hydrocarbons, especially methane, are obtained, because of its high hydrogenation capability [2]. Thus, cobalt and iron are the only relevant catalysts which are currently used commercially. Cobalt is especially advantageous for converting methane-derived syngas because of its high activity and selectivity for linear long-chained hydrocarbons, low deactivation rate, and finally, low activity for water–gas shift (WGS) [3,4]. However, cobalt is more expensive than iron. Indeed, the price of cobalt in the last five years has been within the range of 22–100 $/kg [5].

The activity of a cobalt catalyst depends on the number of exposed Co^0 atoms, the active sites, on the catalyst's surface [6]. Systems with relatively high dispersions are needed in order to maximize the surface availability of Co^0. For this reason, cobalt is typically supported on high surface area carriers with strong interactions (e.g., Al_2O_3, TiO_2). However, thermodynamic studies suggest cobalt nanoparticles lower than 4 nm might be reoxidized by water under FTS reaction conditions [7,8]. Thus, the optimal particle sizes are in the range of 6–10 nm [9]. Even if the particle size, as well as the interaction with the support, are optimized, the majority of the cobalt is locked within the particle instead of being exposed to the surface. Therefore, the incorporation of a second metal, less expensive and with electronic properties close to that of cobalt, could be a possible route to decreasing the total preparation cost of the catalyst. DFT screening was used by Van Helden et al. [10] to identify the alloys which have similar adsorption and electronic properties to the cobalt catalyst. $NiCo_3$, $AlNi_3$, and $SiFe_3$ are suitable cheaper candidates. However, $SiFe_3$ and $AlNi_3$ are quite difficult to produce at the nanoscale level. In contrast, Co–Ni alloys can be easily prepared at different Ni/Co ratios. Ni–Mn bimetallic systems were also investigated for CO_x hydrogenation performance. The authors reported the formation of $NiMnO_3$ for Ni/Mn > 1 and higher activity for CO hydrogenation [11].

Cobalt and nickel have different electronic configurations. Cobalt is a d^7 metal which dissociates CO and stabilizes the vinylic intermediate. This intermediate species is stable in the sp^3 configuration, which favors the chain growth to linear hydrocarbons during FTS [2]. In contrast, nickel (d^8 metal) has a greater electronic back-donation capability. Thus, vinylic intermediates cannot be sufficiently stabilized, thereby favoring the production of light saturated hydrocarbons, especially methane. Therefore, it is of fundamental importance to investigate alloys with different Ni/Co ratios in order to determine the optimal nickel loading for (1) stabilization of the vinylic intermediate, (2) catalyst stability with time on-stream, and (3) retaining high selectivity to longer chained hydrocarbons.

Co-Ni alloys for FTS were investigated by different authors [10,12–22]. Ishihara et al. studied Co-Ni alloys supported on MnO-ZrO_2 [13] and SiO_2 [12,14]. The authors found that the electronic interactions between nickel and cobalt create new adsorption sites, which strengthen the adsorption of hydrogen and enhance the catalytic activity. Further research has suggested that nickel facilitates the reduction of cobalt, shifting it to a lower temperature, as well as increasing the dispersion of cobalt [16,18]. Moreover, Rytter et al. [18] observed that cobalt catalysts with nickel loadings up to 5 wt.% have improved stability because of the suppression of coking through nickel decoration of the cobalt surface. Recently, Lòpez-Tinoco et al. [20] characterized well-controlled nanoparticles consisting of Co-Ni alloys and compared them with a conventional heterogeneous catalyst. TPR-XANES/EXAFS showed that cobalt and nickel have an oxidation state which can be tuned from +2 to 80% metallic. However, well-controlled nanoparticles can currently be prepared in only small amounts, and as such, cannot be easily used for commercial FTS applications. The authors prepared a conventional nickel-cobalt catalyst and found similar activity to Co/Al_2O_3 at steady-state conditions. However, additional investigations are needed to speculate the role of nickel in these bimetallic systems.

In the present work, catalysts with different Ni/Co ratios were prepared and characterized by BET, XRD, ICP, TPR-EXAFS/XANES, and STEM. The activity and the catalytic stability have been evaluated by testing the catalyst in a CSTR reactor for more than 200 h.

2. Results

2.1. Catalyst Characterization

BET surface areas for the prepared catalysts are shown in Table 1. The surface area, pore volume, and pore diameter are similar among all the samples, suggesting that the substitution of cobalt with nickel does not affect the morphological properties. Table 1 also shows the ICP results. The cobalt loading is slightly higher (≈ 30%) than the theoretical value (25%), whereas all Ni/Co ratios are consistent with the nominal values.

Table 1. BET, BJH, and ICP results for the prepared catalysts.

Sample ID	A_s (BET) $[m^2/g]$	V_P (BJH Des) $[cm^3/g]$	D_P (BJH Des) [Å]	Co%	Ni%
25%Co	95.5	0.243	93	30.21	-
25%M—5%Ni-95%Co	92.9	0.226	91	30.49	1.43
25%M—10%Ni-90%Co	96.5	0.227	94	31.58	2.97
25%M—25%Ni-75%Co	96.0	0.236	89	24.86	7.7
25%M—50%Ni-50%Co	91.6	0.237	87	17.4	16.2

XRD patterns for Al_2O_3 and the oxide catalysts are plotted in Figure 1. All the catalysts show the characteristic reflection peaks associated with Co_3O_4 (i.e., 2Θ = 36.8°). No diffraction peaks correlated with nickel compounds are detected for the nickel-promoted catalysts. This suggests nickel is well dispersed, as well as the formation of a Co-Ni solid oxide solution.

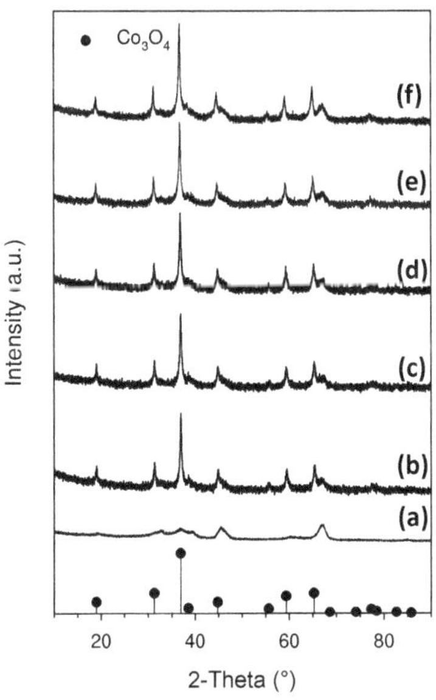

Figure 1. XRD for (**a**) 25%Co/Al_2O_3, (**b**) 25%M (M = 5%Ni-95%Co)/Al_2O_3, (**c**) 25%M (M = 10%Ni-90%Co)/Al_2O_3, (**d**) 25%M (M = 25%Ni-75%Co)/Al_2O_3, and (**e**) 25%M (M = 50%Ni-50%Co)/Al_2O_3.

Figure 2 shows TEM and STEM of reduced 25%M (10%Ni-90%Co)/Al_2O_3. The particle sizes are distributed between 18 and 23 nm, whereas the presence of nickel and cobalt was confirmed by EDS analysis. These two metals are uniformly distributed, confirming the formation of a Co–Ni solid oxide solution, as suggested from XRD results. The Co/Ni weight ratio is close to 10, similar to the theoretical value. The particle sizes for the 25%Co/Al_2O_3 are between 15 and 20 nm (Figure 3). EDS analysis shows areas richer in cobalt (until 55 wt. %), and others poorer in cobalt (10 wt. %); however, the average cobalt loading is close to 28 wt. % over an extended area.

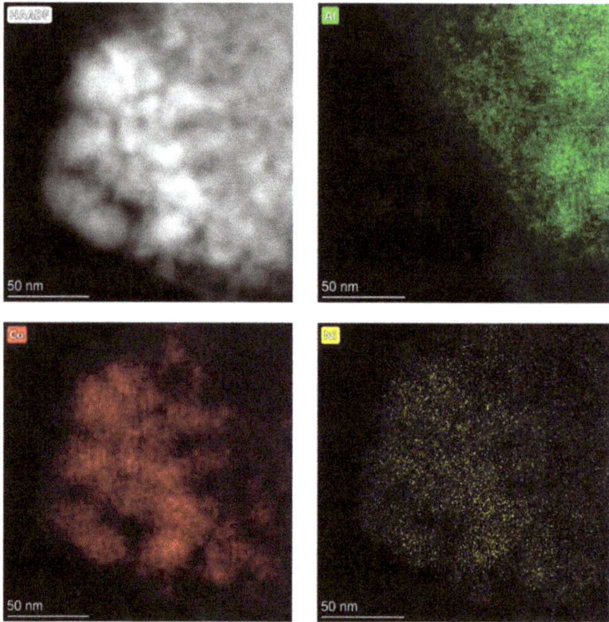

Figure 2. HAADF-STEM image of sample 25%M (10%Ni-90%Co). Elemental mapping legend: (yellow) nickel, (red) cobalt, and (green) aluminum.

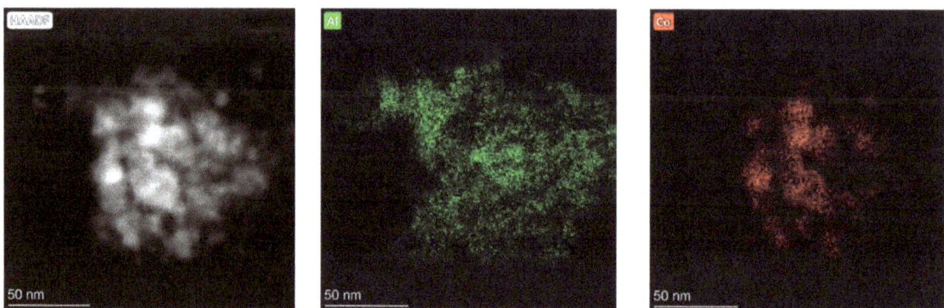

Figure 3. HAADF-STEM image of sample 25%Co. Elemental mapping legend: (green) aluminum and (red) cobalt.

Hydrogen chemisorption with pulse reoxidation results are presented in Table 2. The degree-of-reduction results are similar to those of the TPR-XANES/EXAFS data, as will be shown in the next sections. With increasing Ni/Co ratio, the mixed metal oxides exhibit more facile reduction. There also appears to be a slight increase in average metal diameter with increases in Ni/Co ratio as well. If the traditional approach, designated Method #1, of assuming complete oxidation of reduced metals to their respective oxides is used, then the increase is only slight (i.e., from 11.2 to 15.2 nm). However, if Method #2 is used, the difference is wider (i.e., from 7.0 to 13.5 nm). Method #2 assumes that, during reduction, all Co_3O_4 reduces to CoO, while only a portion of CoO reduces to Co^0 (and a fraction of NiO reduces to Ni^0). Thus, during the reoxidation with O_2 pulses, the Ni^0 and Co^0

oxidize to NiO and CoO, and all CoO (including CoO obtained from Co^0 oxidation as well as the CoO previously resulting from merely partial reduction of Co_3O_4) oxidizes to Co_3O_4.

Table 2. H_2 chemisorption and pulse O_2 titration.

μmol H_2 desorbed/g_{cat}	Uncorr. % Disp.	Uncorr. Diam. (nm)	O_2 Uptake (mol/g_{cat})	* % Red.	** % Red.	* Corr. % Disp.	** Corr. % Disp.	* Corr. Diam. (nm)	** Corr. Diam. (nm)
25%Co/Al_2O_3									
91.3	4.3	24	1324	46.8	29.1	9.2	14.8	11.2	7.0
25%M (95%Co-5%Ni)/Al_2O_3									
104.0	4.9	21.1	1317	47.3	30.4	10.4	16.1	10.0	6.4
25%M (90%Co-10%Ni)/Al_2O_3									
92.5	4.4	24	1495	54.6	40.5	8.0	10.8	12.9	9.6
25%M (75%Co-25%Ni)/Al_2O_3									
94.6	4.5	23.2	1563	59.9	48.7	7.5	9.2	13.9	11.3
25%M (50%Co-50%Ni)/Al_2O_3									
94.7	4.5	23.1	1594	65.7	58.5	6.8	7.6	15.2	13.5

* Method #1 assuming Ni^0 oxidizes to NiO and Co^0 oxidizes to Co_3O_4. ** Method #2 assuming all Co_3O_4 reduced to CoO and some NiO and CoO reduced to Ni^0 and Co^0. During oxidation, then, the Ni^0 and Co^0 oxidize to NiO and CoO, and all CoO oxidizes to Co_3O_4.

2.2. *Cobalt Reducibility*

2.2.1. H_2 TPR-XANES

Figures 4 and 5 display H_2 TPR-XANES spectra for the prepared samples at the Co K-edge as a function of increasing Ni/Co ratio using two different perspectives. The perspective of Figure 4 is versus temperature, whereas the perspective of Figure 5 is that of photon energy. XANES snapshots at the point of 100% Co_3O_4, 100% CoO, as well as the final spectrum at the point of maximum reduction to Co^0, are shown in Figure 6. Cobalt oxides reduce to metallic compounds in two steps: (I) $Co_3O_4 + H_2 = 3CoO + H_2O$ and (II) $3CoO + 3H_2 = 3Co^0 + 3H_2O$. Figure 6 shows that with increasing Ni/Co ratio, the reduction of cobalt oxides systematically moves to lower temperature.

Linear combination XANES fittings at the Co K-edge are provided in Figure 7. For all of the catalysts, the initial spectrum at 25 °C resembles that of Co_3O_4. The point of 50%Co_3O_4/50%CoO was reached at 339 °C (Ni/Co = 0/100), 308 °C (Ni/Co = 5/95), 294 °C (Ni/Co = 10/90), 273 °C (Ni/Co = 25/75), and 249 °C (Ni/Co = 50/50). Thus, up to a Δ 90 °C decrease in reduction temperature was achieved. The point of 100%CoO was obtained at 400 °C (Ni/Co = 0/100), 330 °C (Ni/Co = 5/95), 330 °C (Ni/Co = 10/90), 305 °C (Ni/Co = 25/75), and 278 °C (Ni/Co = 50/50). Thus, up to a Δ 122 °C decrease in reduction temperature was obtained in converting Co_3O_4 to CoO by increasing the Ni/Co ratio.

The point of 50%CoO/50%Co^0 was attained at 553 °C (Ni/Co = 0/100), 473 °C (Ni/Co = 5/95), 473 °C (Ni/Co = 10/90), 437 °C (Ni/Co = 25/75), and 386 °C (Ni/Co = 50/50). Thus, up to a Δ 167 °C decrease in reduction temperature was achieved by doping Co with Ni. It is evident that, unlike the Group 10 metal Pt, where substantial shifts in the reduction of Co oxides were observed with minute amounts of Pt, significantly higher quantities of Ni are required to achieve the same level of reduction. For example, just 0.5%Pt was able to facilitate a decrease of 194 °C for the Co_3O_4 to CoO transition, and a decrease of 120 °C from CoO to Co^0 [23], which is < 1/40th of the atomic amount. In comparison with the TPR profile of CoO during reduction of undoped 25%Co/Al_2O_3, this is likely in part due to the strong interactions between Ni oxides and alumina support.

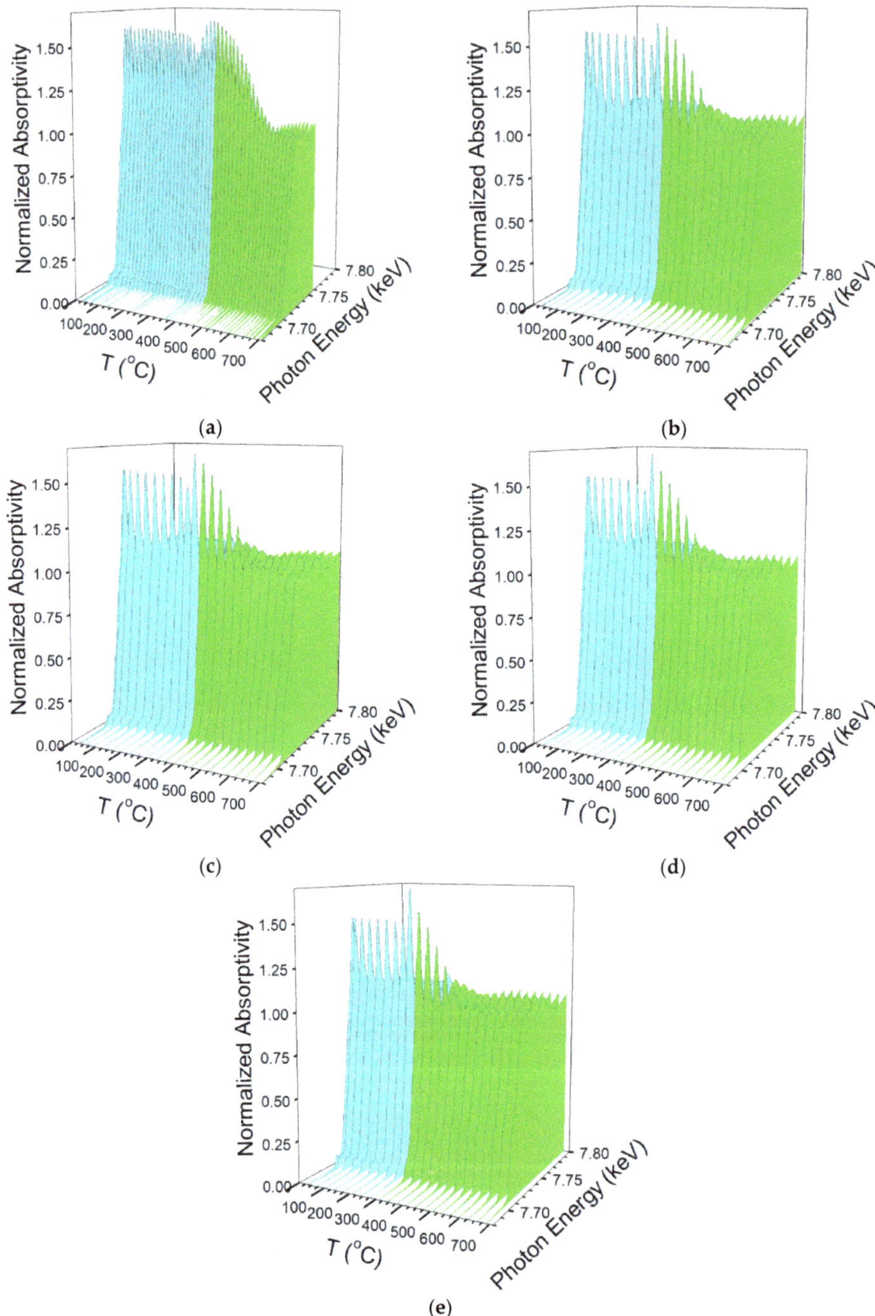

Figure 4. H_2-TPR-XANES spectra at the Co K-edge of (**a**) 25%Co/Al_2O_3, (**b**) 25%M (M = 5%Ni-95%Co)/Al_2O_3, (**c**) 25%M (M = 10%Ni-90%Co)/Al_2O_3, (**d**) 25%M (M = 25%Ni-75%Co)/Al_2O_3, and (**e**) 25%M (M = 50%Ni-50%Co)/Al_2O_3.

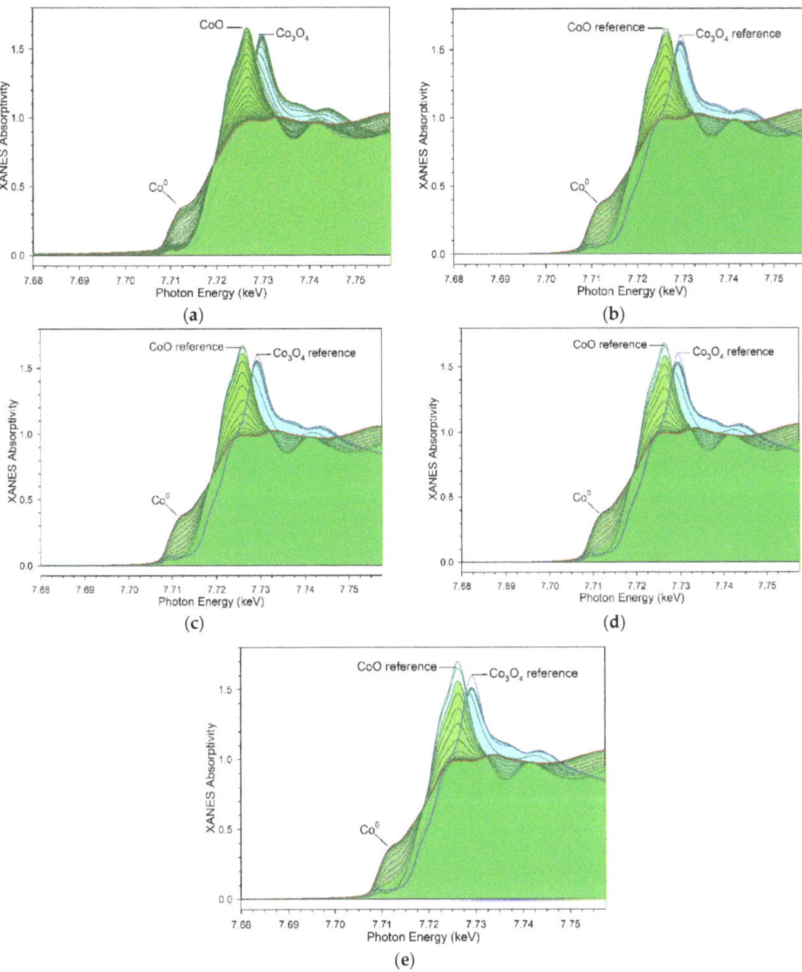

Figure 5. H_2-TPR-XANES spectra (XY view) at the Co K-edge of (**a**) 25%Co/Al_2O_3, (**b**) 25%M (M = 5%Ni-95%Co)/Al_2O_3, (**c**) 25%M (M = 10%Ni-90%Co)/Al_2O_3, (**d**) 25%M (M = 25%Ni-75%Co)/Al_2O_3, and (**e**) 25%M (M = 50%Ni-50%Co)/Al_2O_3.

2.2.2. H_2 TPR-EXAFS

TPR-EXAFS spectra at the Co K-edge are shown in Figures 8 and 9, including a plot that highlights the temperature (Figure 8), and a plot that emphasizes differences as a function of distance from the absorber (Figure 9). The initial cyan spectrum of each TPR-EXAFS profile represents Co_3O_4. It can be differentiated from CoO (i.e., the first green spectrum) by the fact that the Co–O peak is significantly more intense, while the Co–Co coordination peak is broadened due to the presence of additional oxygen atoms (Figure 9). Following the transition to CoO, the CoO slowly converts to Co^0, resulting in a final well-resolved peak for Co–Co metal coordination. In a manner similar to Pt and Ru promoters, Ni facilitates both steps of reduction; however, as mentioned previously, on an atomically equivalent basis, Ni is far less effective than either Pt or Ru [23]. It differs from Re, which only catalyzes the second step, CoO reduction to Co^0. Unlike Pt and Ru oxides, which reduce at low temperatures, Re oxide was observed to reduce at a similar range as Co_3O_4 reduction to CoO (i.e., 300–350 °C), and it was supposed that a reduced form was necessary in order to facilitate CoO reduction through a H_2 dissociation and spillover mechanism [23]. Thus, if a hydrogen dissociation and spillover mechanism operated for the case of Ni, one would expect that NiO should reduce to Ni^0 prior to the reduction of Co oxides. Based on the Ni X-ray absorption spectroscopy results, this does not seem to be the case.

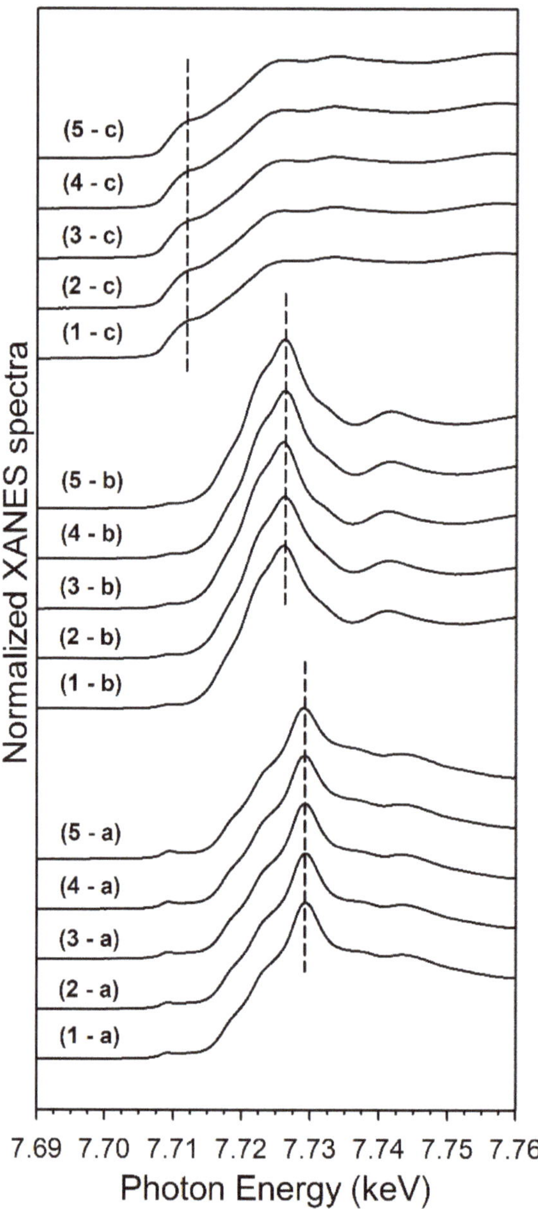

Figure 6. Co K-edge XANES spectra: (**a**) the initial point consisting of primarily Co_3O_4; (**b**) the point of maximum CoO content; and (**c**) the final spectrum consisting of primarily Co^0 for (**1**) 25%Co/Al_2O_3, (**2**) 25%M (M = 5%Ni-95%Co)/Al_2O_3, (**3**) 25%M (M = 10%Ni-90%Co)/Al_2O_3, (**4**) 25%M (M = 25%Ni-75%Co)/Al_2O_3, and (**5**) 25%M (M = 50%Ni-50%Co)/Al_2O_3.

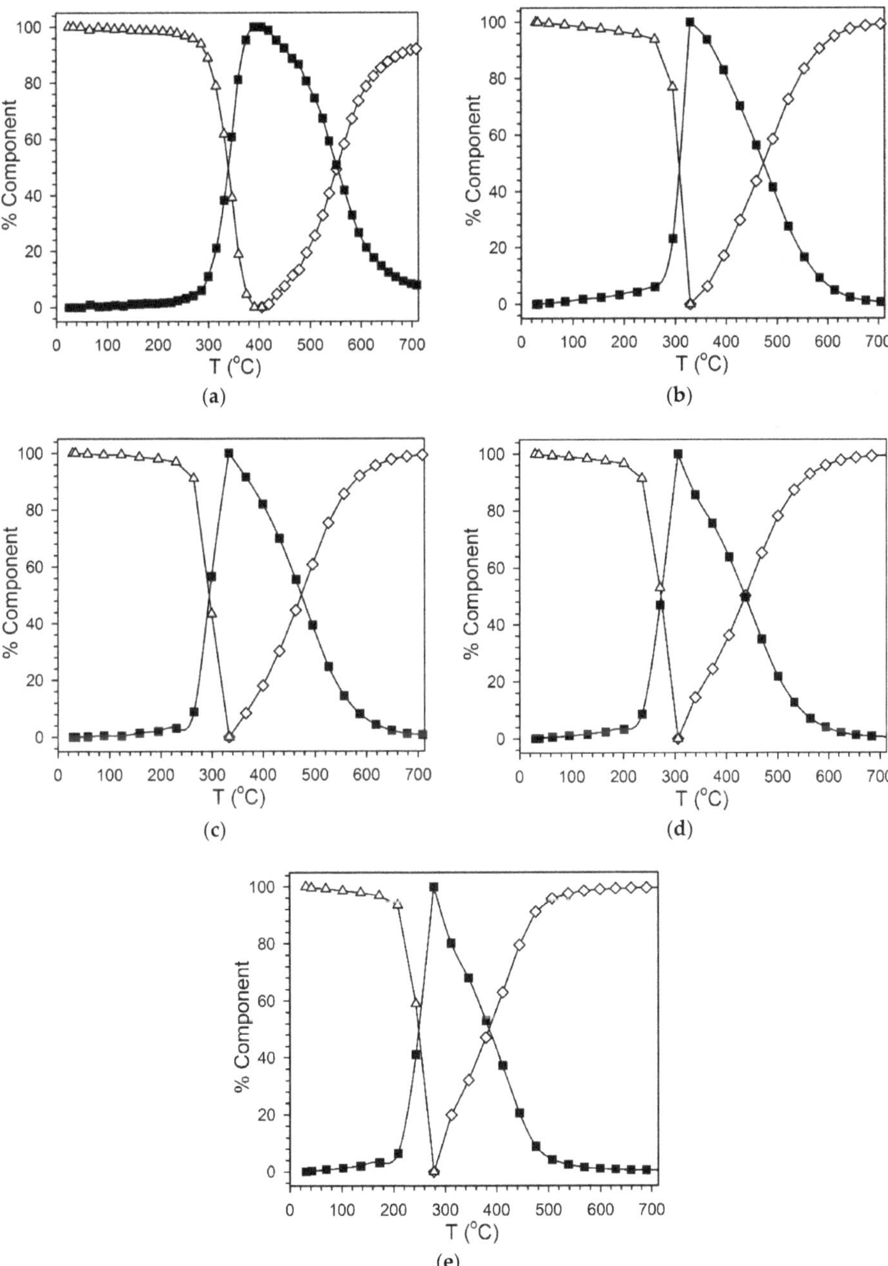

Figure 7. LC fittings of H_2-TPR-XANES spectra at the Co K-edge of (**a**) 25%Co/Al_2O_3, (**b**) 25%M (M = 5%Ni-95%Co)/Al_2O_3, (**c**) 25%M (M = 10%Ni-90%Co)/Al_2O_3, (**d**) 25%M (M = 25%Ni-75%Co)/Al_2O_3, and (**e**) 25%M (M = 50%Ni-50%Co)/Al_2O_3.

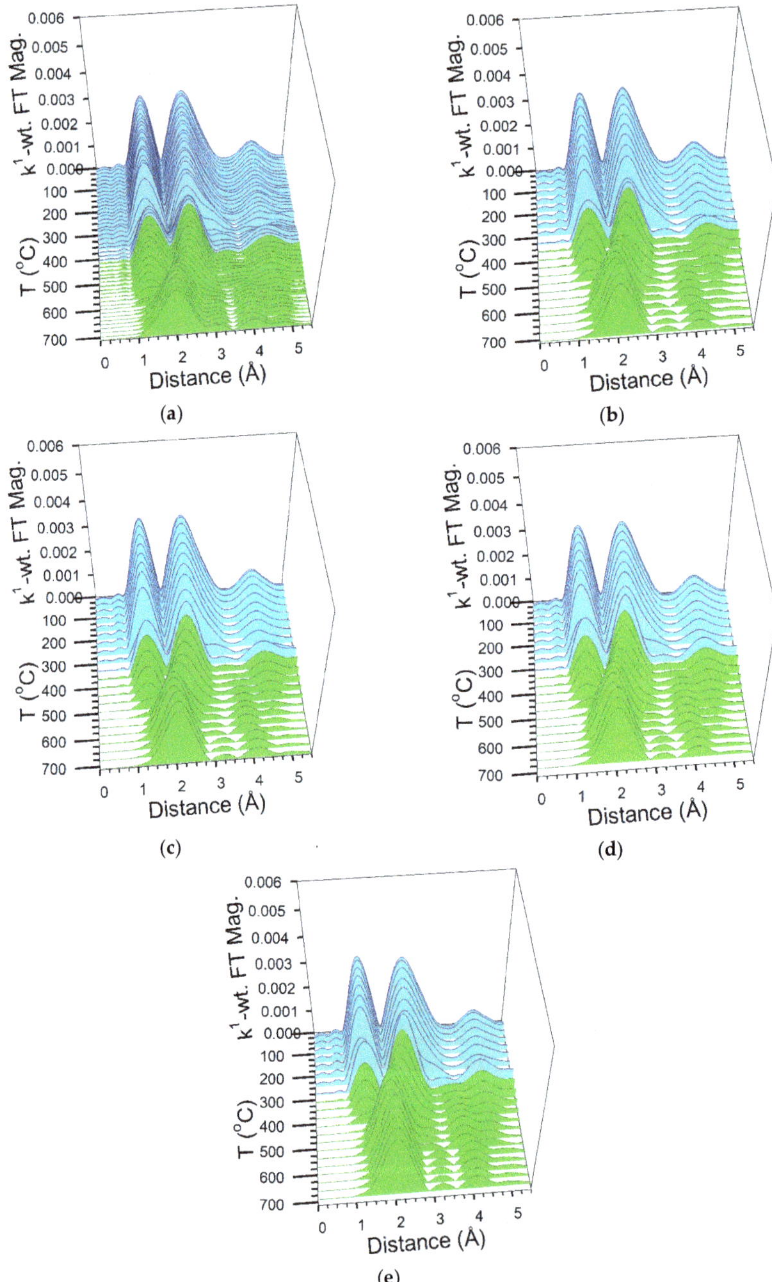

Figure 8. H_2-TPR-EXAFS spectra at the Co K-edge of (**a**) 25%Co/Al_2O_3, (**b**) 25%M (M = 5%Ni-95%Co)/Al_2O_3, (**c**) 25%M (M = 10%Ni-90%Co)/Al_2O_3, (**d**) 25%M (M = 25%Ni-75%Co)/Al_2O_3, and (**e**) 25%M (M = 50%Ni-50%Co)/Al_2O_3. (Cyan) is reduction of Co_3O_4 to CoO, and (Green) CoO to Co^0.

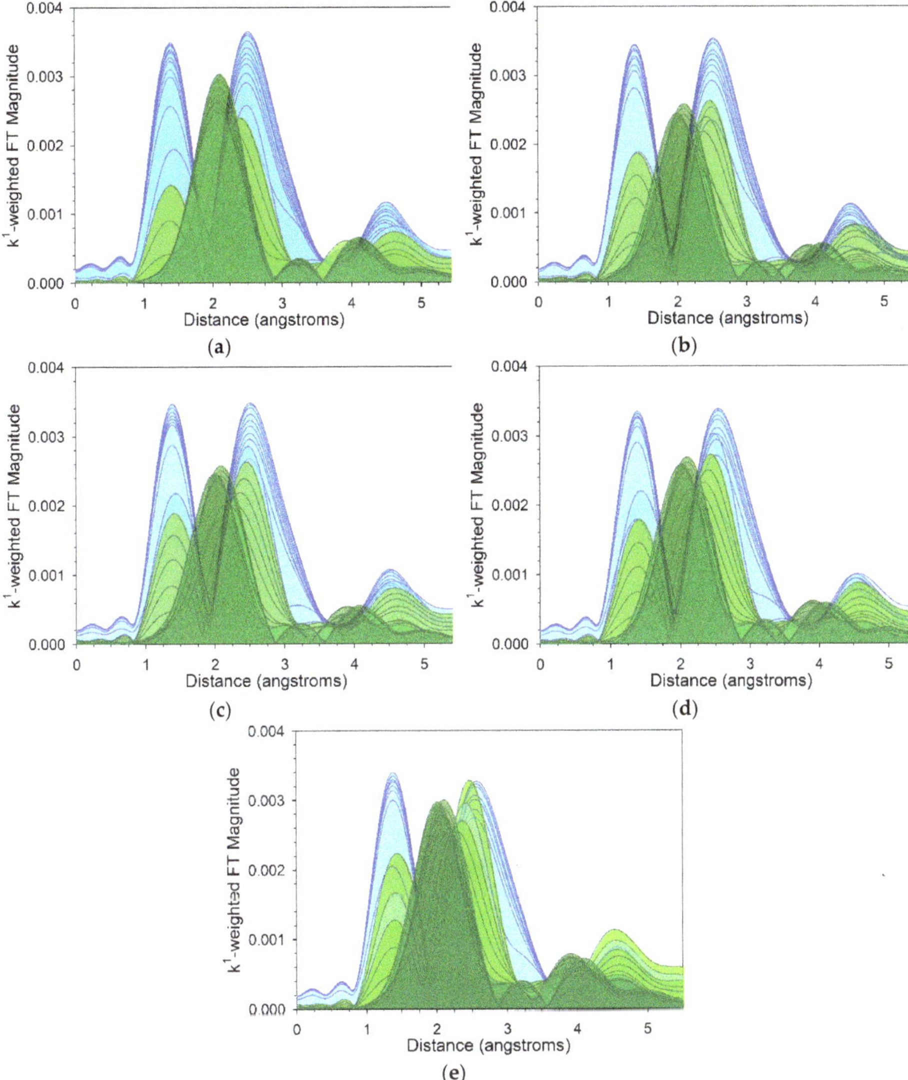

Figure 9. H_2-TPR-EXAFS spectra (XY view) at the Co K-edge of (**a**) 25%Co/Al_2O_3, (**b**) 25%M (M = 5%Ni-95%Co)/Al_2O_3, (**c**) 25%M (M = 10%Ni-90%Co)/Al_2O_3, (**d**) 25%M (M = 25%Ni-75%Co)/Al_2O_3, and (**e**) 25%M (M = 50%Ni-50%Co)/Al_2O_3. (Cyan) is reduction of Co_3O_4 to CoO, and (Green) CoO to Co^0.

EXAFS snapshots of the different chemical species (e.g., Co_3O_4, CoO, and Co^0), observed during the TPR trajectory, are provided in Figure 10. Structural similarities are evidenced by the common peak positions for Co-O and Co-Co coordination. The only differences among the samples are in the temperatures at which the CoO and Co^0 evolve, as previously discussed in the TPR-XANES section.

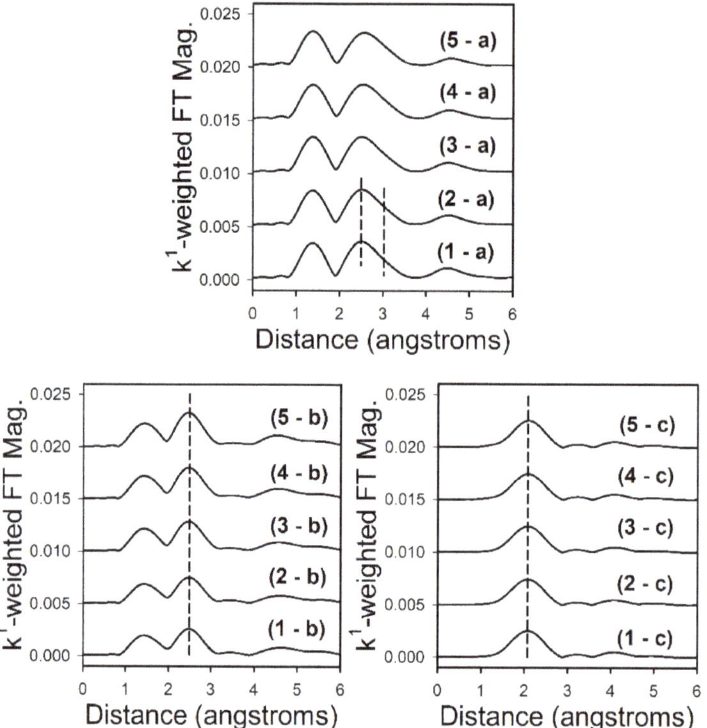

Figure 10. Co K-edge EXAFS spectra of (**a**) the initial point consisting of primarily Co_3O_4, (**b**) the point of maximum CoO content, and (**c**) the final spectrum consisting of primarily Co^0 for (**1**) 25%Co/Al_2O_3, (**2**) 25%M (M = 5%Ni-95%Co)/Al_2O_3, (**3**) 25%M (M = 10%Ni-90%Co)/Al_2O_3, (**4**) 25%M (M = 25%Ni-75%Co)/Al_2O_3, (**5**) 25%M (M = 50%Ni-50%Co)/Al_2O_3.

2.3. *Nickel Reducibility*

2.3.1. H_2 TPR-XANES

TPR-XANES spectra in Figures 11 and 12 at the Ni K-edge reveal that, initially, Ni oxide is associated with Co_3O_4 (see spectra in cyan color), and this Ni oxide subsequently undergoes a change in electronic structure to a form of Ni oxide associated with CoO (initial green spectrum). This change is best observed by examining the head-on spectra of Figure 13, as well as the XANES snapshots shown in Figure 14. Let us refer to this transition as step one. Step two is further reduction of Ni^{2+} to Ni^0 (dark green spectra in Figure 12).

Figure 14 provides quantitative information regarding the Ni species present along the TPR trajectory. In step one, the point of 50%Ni^{2+}-Co_3O_4 / 50%Ni^{2+}-CoO was reached at 333 °C (Ni/Co = 5/95), 286 °C (Ni/Co = 10/90), 261 °C (Ni/Co = 25/75), and 234 °C (Ni/Co = 50/50). Thus, up to a Δ 99 °C decrease in reduction temperature was achieved (compared to a Δ of 60 °C for the cobalt system over the same range of loading), and the temperatures match well with those of the Co_3O_4 to CoO transitions described earlier at the 50% point of conversion, especially at higher Ni/Co ratios. The point of 100%Ni^{2+}-CoO was obtained at 362 °C (Ni/Co = 5/95), 332 °C (Ni/Co = 10/90), 305 °C (Ni/Co = 25/75), and 278 °C (Ni/Co = 50/50). Thus, up to a Δ 84 °C decrease in reduction temperature was obtained in converting 100%Ni^{2+}-Co_3O_4 to 100%Ni^{2+}-CoO (compared to a Δ of 52 °C for the cobalt system over the same range of loading). Once again, the temperatures match quite well with those of the Co_3O_4 to

CoO transitions described earlier for Co K-edge data, and match nearly perfectly at Ni/Co loadings of 10%, 25%, and 50%.

Continuing, the point of 50%Ni^{2+}-CoO/50%Ni^0 was achieved at 503 °C (Ni/Co = 5/95), 482 °C (Ni/Co = 10/90), 458 °C (Ni/Co = 25/75), and 412 °C (Ni/Co = 50/50). Thus, up to a Δ 91 °C decrease in reduction temperature was achieved by doping Co with Ni (compared to a Δ of 87 °C for the cobalt system over the same range of loading).

The similarities in temperature ranges between Co_3O_4 to CoO and the Ni^{2+}-Co_3O_4 to Ni^{2+}-CoO transitions, as well those between CoO to Co^0 and Ni^{2+}-CoO to Ni^0-Co^0 transitions, suggest that the effect may not be simply a H_2 dissociation and spillover mechanism, but rather a chemical effect due to intimate contact between Ni and Co in both oxide (e.g., solid solution) and metallic (e.g., alloy) phases, throughout the TPR trajectory.

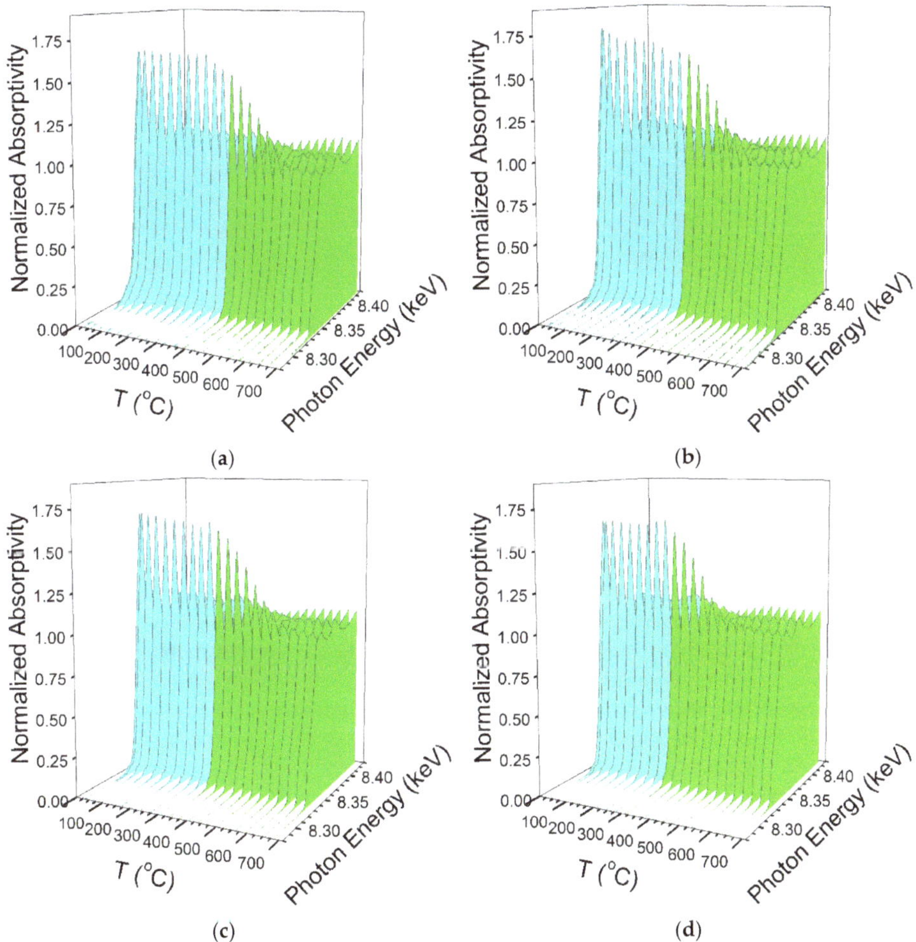

Figure 11. H_2-TPR-XANES spectra at the Ni K-edges of (**a**) 25%M (M = 5%Ni-95%Co)/Al_2O_3, (**b**) 25%M (M = 10%Ni-90%Co)/Al_2O_3, (**c**) 25%M (M = 25%Ni-75%Co)/Al_2O_3, and (**d**) 25%M (M = 50%Ni-50%Co)/Al_2O_3. (Cyan) is Ni^{2+} (e.g., NiO) associated with cobalt oxides during reduction of Co_3O_4 to CoO. (Green) is reduction of Ni^{2+} to Ni^0 when NiO reduction is associated with CoO reduction to Co^0.

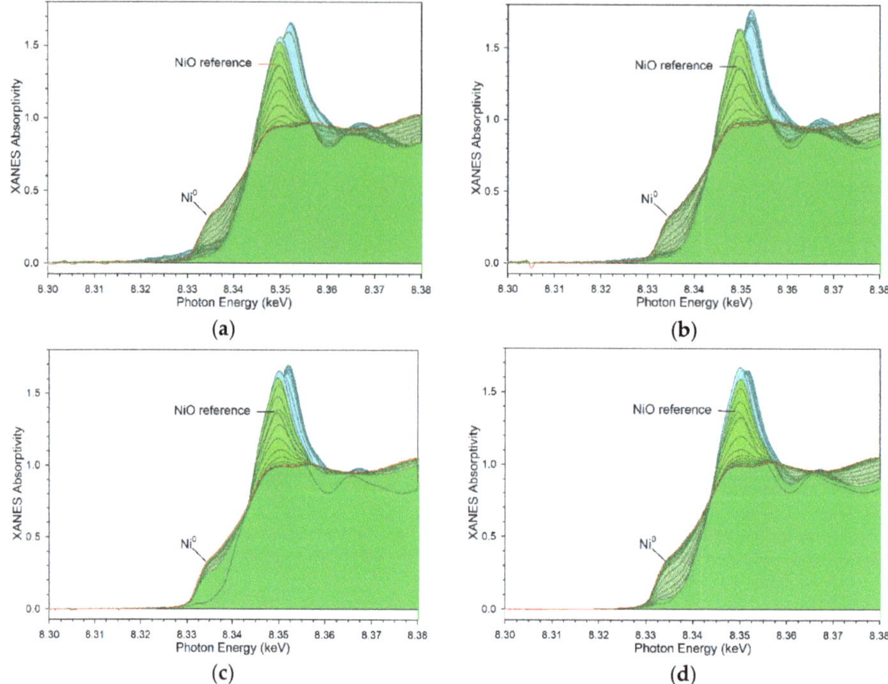

Figure 12. H_2-TPR-XANES spectra at the Ni K-edges of (**a**) 25%M (M = 5%Ni-95%Co)/Al_2O_3, (**b**) 25%M (M = 10%Ni-90%Co)/Al_2O_3, (**c**) 25%M (M = 25%Ni-75%Co)/Al_2O_3, and (**d**) 25%M (M = 50%Ni-50%Co)/Al_2O_3. (Cyan) is Ni^{2+} (e.g., NiO) associated with cobalt oxides during reduction of Co_3O_4 to CoO. (Green) is reduction of Ni^{2+} to Ni^0 when NiO reduction is associated with CoO reduction to Co^0.

2.3.2. H_2 TPR-EXAFS

Comparing the TPR-EXAFS spectra of Figure 15 with Figure 8, Figure 16 with Figure 9, and Figure 17 with Figure 10, it is remarkable how closely the behavior of Ni resembles that of Co. The initial cyan spectra of Figures 15 and 16 reveal that the first Ni–O peak is more intense than that of the green Ni–O peak, suggesting greater coordination to oxygen, while the second peak for Ni–Ni coordination in the initial cyan spectra is more broadened compared to the initial green spectra, consistent with greater oxygen content forcing Ni atoms apart. Figure 17 highlights the differences between the local atomic structures of different species along the TPR profile (e.g., Ni^{2+} associated with Co_3O_4; Ni^{2+} associated with CoO; and Ni^0 associated with Co^0).

Figures 18 and 19, as well as Table 3, show EXAFS fittings for the Co K-edge and Ni K-edge data following TPR-EXAFS and cooling to ambient temperature. A simple model was developed previously [20] and applied here, wherein metal coordination to Ni (whether the core atom was Co or Ni) was given as a fraction, X, of metal-cobalt coordination. Using this approach, excellent fittings with low r-factors were obtained. Co-Ni alloy formation is consistent with EXAFS fitting, but it cannot be considered proven, since Co and Ni are too closely related in size.

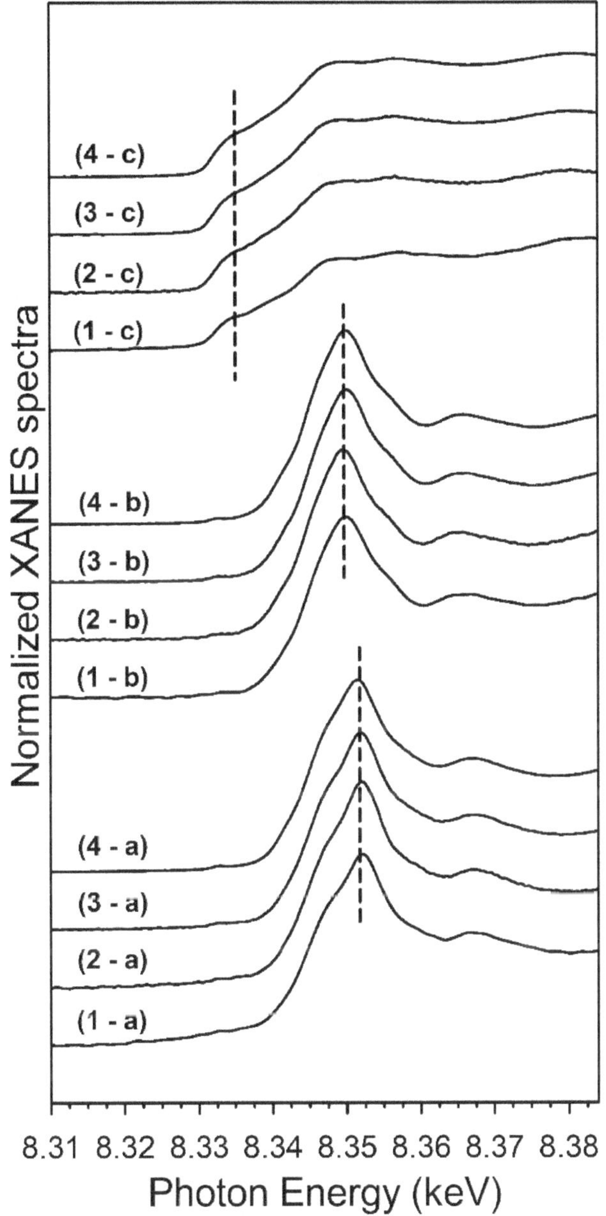

Figure 13. Ni K-edge XANES spectra of (**a**) the initial point consisting of primarily NiO associated with Co_3O_4, (**b**) the point consisting of primarily NiO associated with CoO, and (**c**) the final spectrum consisting of primarily Ni^0 for (**2**) 25%M (M = 5%Ni-95%Co)/Al_2O_3, (**3**) 25%M (M = 10%Ni-90%Co)/Al_2O_3, (**4**) 25%M (M = 25%Ni-75%Co)/Al_2O_3.

Table 3. Results of EXAFS fitting* for data acquired near the Co and Ni K edges for catalysts following TPR-EXAFS after cooling. The fitting ranges were $\Delta k = 3–10$ $Å^{-1}$ and $\Delta R = 1.2–2.8$ Å. $*S_0{}^2$ set to 0.90. Mixing parameter fixed to nominal value.

Sample Description	N Co–Co Metal	R Co–Co (Å) Metal	N Co-Ni Metal	R Co-Ni (Å) Metal	N Ni-Ni Metal	R Ni-Ni (Å) Metal	N Ni-Co Metal	R Ni-Co (Å) Metal	e_0 (eV)	σ^2 ($Å^2$)	r-factor
100Co	9.9	2.489 (0.003)	-	-	-	-	-	-	6.37 (0.45)	0.00731 (0.00043)	0.0010
5Ni:95Co	9.9 (0.78)	2.489 (0.0053)	0.50 (0.04)	2.481 (0.0053)	0.32 (0.03)	2.472 (0.0053)	6.3 (0.59)	2.481 (0.0053)	6.32 (0.813)	0.00787 (0.00076)	0.014
10Ni:90Co	9.4 (0.48)	2.492 (0.0034)	0.94 (0.05)	2.483 (0.0034)	0.75 (0.04)	2.475 (0.0034)	7.5 (0.41)	2.483 (0.0034)	6.56 (0.527)	0.00777 (0.00048)	0.0035
25Ni:75Co	8.0 (0.31)	2.491 (0.0026)	2.0 (0.08)	2.482 (0.0026)	1.9 (0.08)	2.474 (0.0026)	7.4 (0.32)	2.482 (0.0026)	6.71 (0.401)	0.00719 (0.00036)	0.0023
50Ni:50Co	6.5 (0.36)	2.491 (0.0035)	3.3 (0.18)	2.483 (0.0035)	3.2 (0.18)	2.475 (0.0035)	6.4 (0.36)	2.483 (0.0035)	–4.39 (0.613)	0.00659 (0.00049)	0.0034

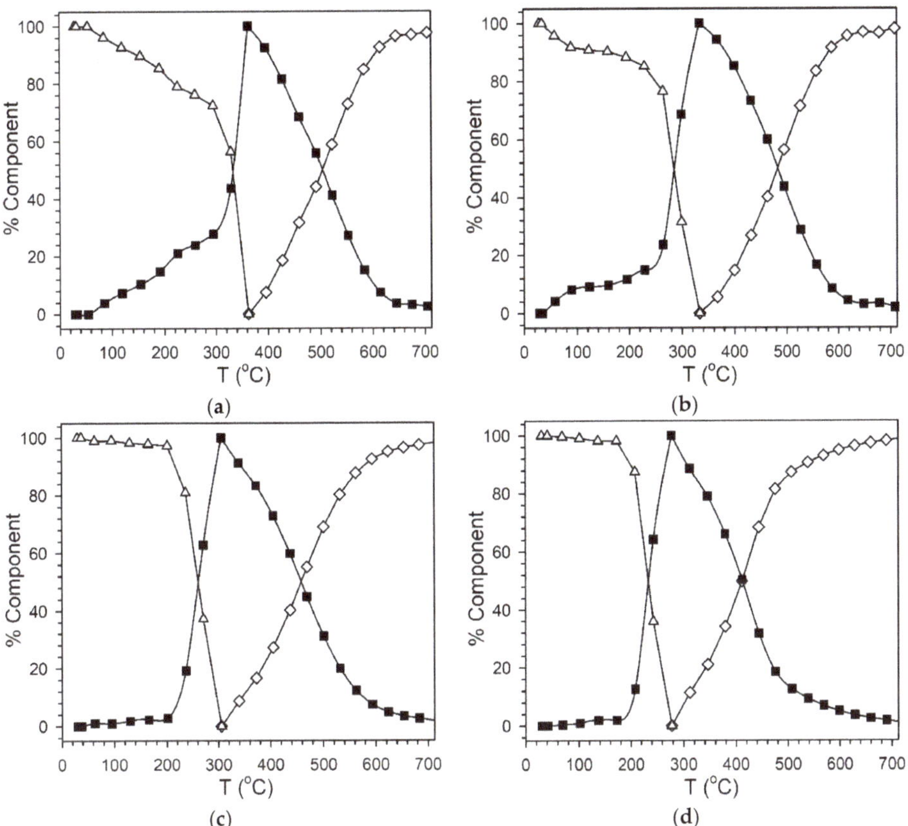

Figure 14. LC fittings of H_2-TPR-XANES spectra at the Ni K-edge of (**a**) 25%M (M = 5%Ni-95%Co)/Al_2O_3, (**b**) 25%M (M = 10%Ni-90%Co)/Al_2O_3, (**c**) 25%M (M = 25%Ni-75%Co)/Al_2O_3, and (**d**) 25%M (M = 50%Ni-50%Co)/Al_2O_3. Legend: △ Ni^{2+} associated with Co_3O_4, ■ Ni^{2+} associated with CoO, and ◇ Ni^0 associated with Co^0.

2.4. Catalytic Activity

CO conversion in the first 200 h is shown in Figure 20. The CO conversion for the unpromoted catalyst has an initial value of 42.5%, and then it slightly decreases in the first few hours until a steady-state value of 39.5% was reached. This trend is typical for cobalt-based catalysts. In contrast, CO conversion progressively increases for all nickel-promoted catalysts. 5%Ni-95%Co and 10%Ni-90% exhibit similar CO conversion trends, as the former's initial value is close to 20%, and it continuously increases, reaching 34%, whereas 25%Ni-75Co has a higher initial CO conversion (29%), and it reaches a steady-state value of 36%.

The evolutions of selectivities (CH_4, CO_2, C_2-C_4, and C_{5+}) with T.o.S. (time on-stream) are shown in Figure 21. CH_4 selectivity and C_{5+} selectivities are stable at 7.6% and 80.7%, respectively, for the unpromoted catalyst. The addition of nickel increases initial CH_4 selectivity, whereas it decreases initial C_{5+} selectivity. Fortunately, CH_4 selectivity for the nickel-promoted catalyst slowly decreased with T.o.S. For example, the initial CH_4 selectivity for 25%M—25%Ni-75%Co is 13.4%, but it reaches 9.5% after 200 h. Thus, the difference compared to Co/Al_2O_3 decreases from 5.8% (absolute) to just 1.9%.

The olefin/paraffin ratio for unpromoted and nickel-promoted catalysts decreases with increasing carbon number, starting with ethylene species and moving upward (Figure 22). This trend is typically observed for cobalt based catalysts. However, even if the trend with the carbon number is similar among the different catalysts, the olefin content decreases by increasing the nickel content. Indeed, the olefin/paraffin ratio for C_3 species is 2.5 for 25%Co, while it is only 1 for the 25%M—75%Co-25%Ni catalyst. At higher carbon numbers, the differences among the different Ni/Co catalysts are not so pronounced because of the tendency of olefins to readsorb on active sites [24]. The high hydrogenation capability of nickel promoted samples is also observed by a slight decrease in chain growth probability. Indeed $\alpha_{c5\text{-}c16}$ slightly decreases from 0.85 for 25%Co to 0.83 for 25%M—25%Ni-75%Co.

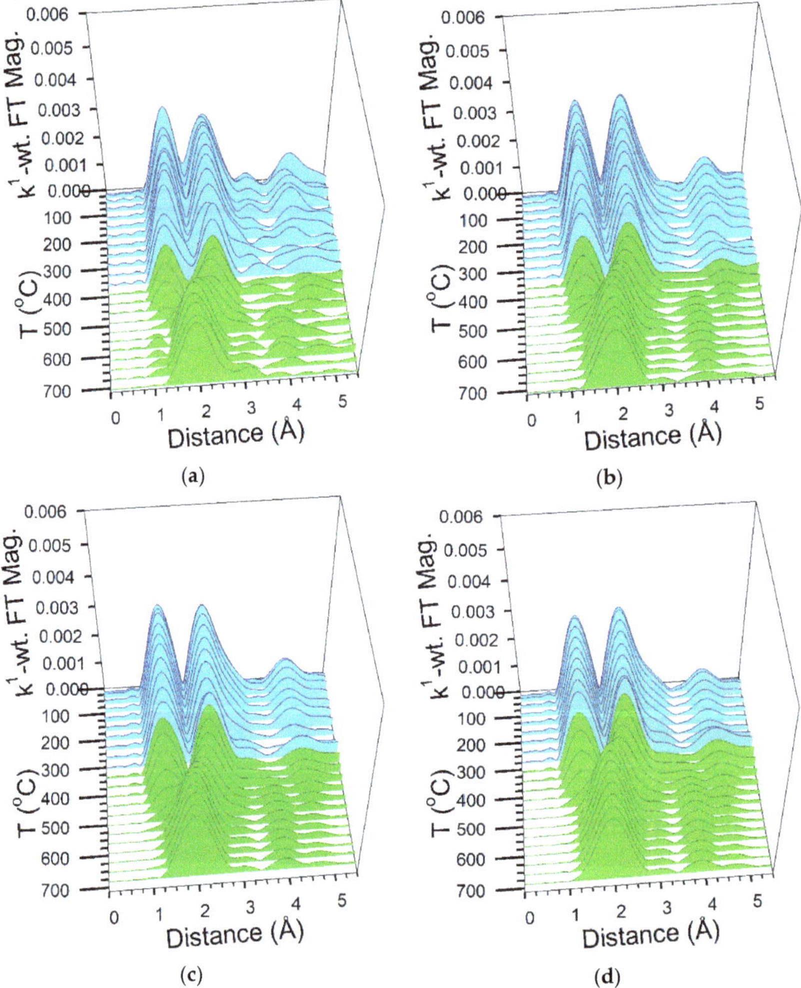

Figure 15. H_2-TPR-EXAFS spectra at the Ni K-edges of (**a**) 25%M (M = 5%Ni-95%Co)/Al_2O_3, (**b**) 25%M (M = 10%Ni-90%Co)/Al_2O_3, (**c**) 25%M (M = 25%Ni-75%Co)/Al_2O_3, and (**d**) 25%M (M = 50%Ni-50%Co)/Al_2O_3. (Cyan) is Ni^{2+} (e.g., NiO) associated with cobalt oxides during reduction of Co_3O_4 to CoO. (Green) is reduction of Ni^{2+} to Ni^0 when NiO reduction is associated with CoO reduction to Co^0.

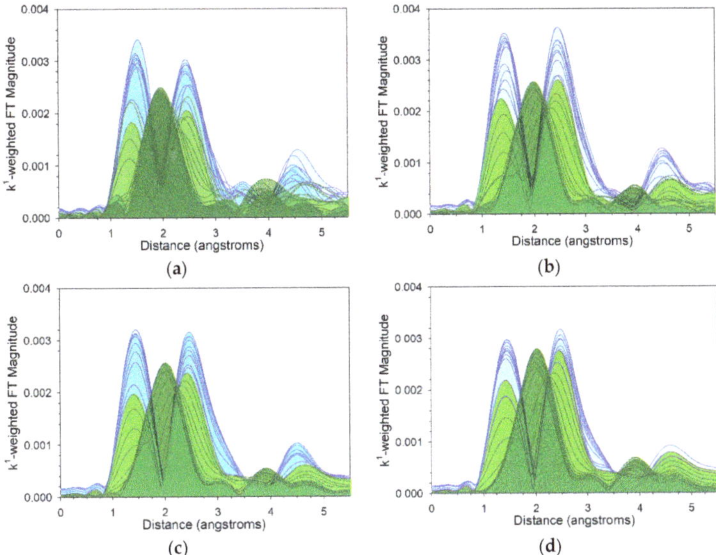

Figure 16. H_2-TPR-EXAFS spectra at the Ni K-edges of (**a**) 25%M (M = 5%Ni-95%Co)/Al_2O_3, (**b**) 25%M (M = 10%Ni-90%Co)/Al_2O_3, (**c**) 25%M (M = 25%Ni-75%Co)/Al_2O_3, and (**d**) 25%M (M = 50%Ni-50%Co)/Al_2O_3. (Cyan) is Ni^{2+} (e.g., NiO) associated with cobalt oxides during reduction of Co_3O_4 to CoO. (Green) is reduction of Ni^{2+} to Ni^0 when NiO reduction is associated with CoO reduction to Co^0.

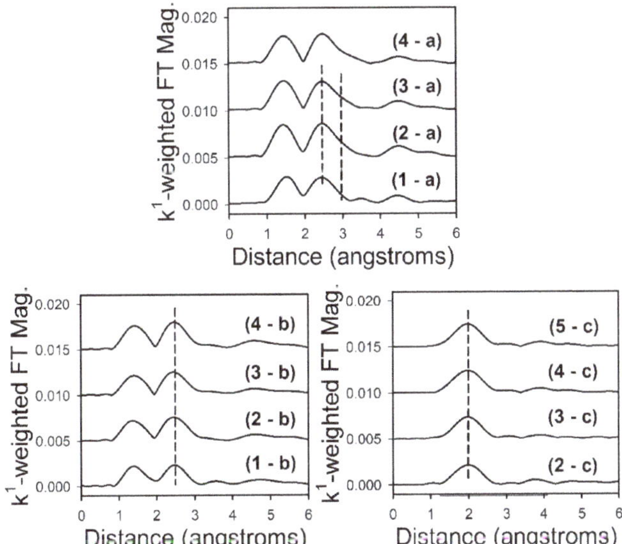

Figure 17. Ni K-edge EXAFS spectra: (**a**) the initial point consisting of primarily NiO associated with Co_3O_4; (**b**) the point consisting of primarily NiO associated with CoO; and (**c**) the final spectrum consisting of primarily Ni^0 for (**2**) 25%M (M = 5%Ni-95%Co)/Al_2O_3, (**3**) 25%M (M = 10%Ni-90%Co)/Al_2O_3, (**4**) 25%M (M = 25%Ni-75%Co)/Al_2O_3, and (**5**) 25%M (M = 50%Ni-50%Co)/Al_2O_3.

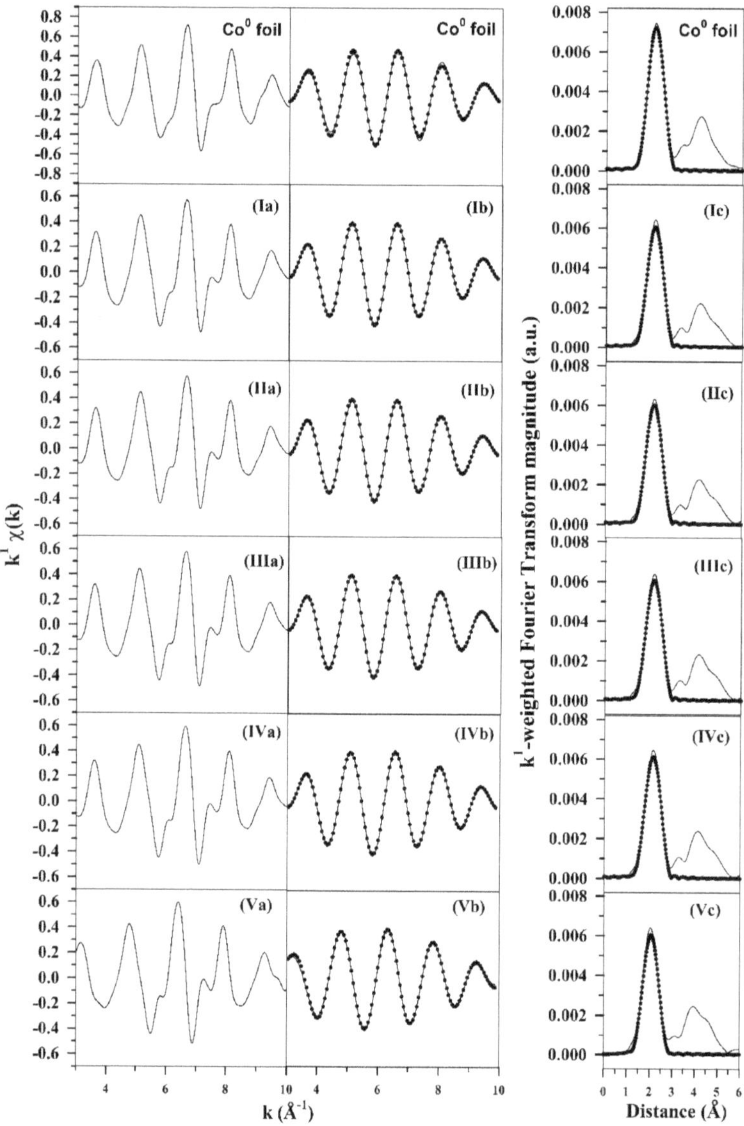

Figure 18. EXAFS fittings for Co K-edge data, including (**a**) raw k^1-weighted χ(k) data; (**b**) (solid line) filtered k^1-weighted χ(k) data and (filled circles) results of the fittings; (**c**) (solid line) raw k^1-weighted Fourier transform magnitude and (solid line) filtered k^1-weighted Fourier transform magnitude; and (filled circles) results of the fittings for Co^0 foil, (**I**) 25%Co/Al_2O_3, (**II**) 25%M (M = 5%Ni-95%Co)/Al_2O_3, (**III**) 25%M (M = 10%Ni-90%Co)/Al_2O_3, (**IV**) 25%M (M = 25%Ni-75%Co)/Al_2O_3, and (**V**) 25%M (M = 50%Ni-50%Co)/Al_2O_3.

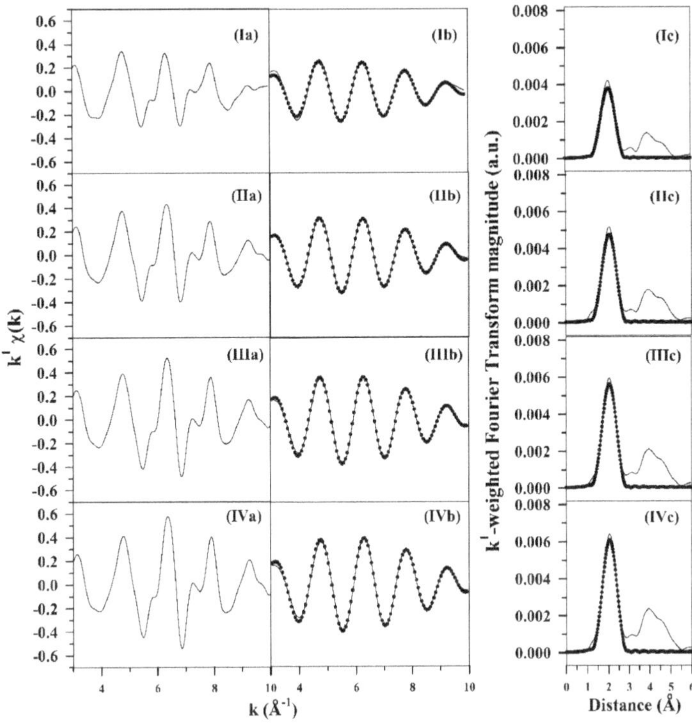

Figure 19. EXAFS fittings for Ni K-edge data, including (**a**) raw k^1-weighted (k) data; (**b**) (solid line) filtered k^1-weighted (k) data and (filled circles) results of the fittings; (**c**) (solid line) raw k^1-weighted Fourier transform magnitude and (solid line) filtered k^1-weighted Fourier transform magnitude; and (filled circles) results of the fittings for (**I**) 25%M (M = 5%Ni-95%Co)/Al_2O_3, (**II**) 25%M (M = 10%Ni-90%Co)/Al_2O_3, (**III**) 25%M (M = 25%Ni-75%Co)/Al_2O_3, and (**IV**) 25%M (M = 50%Ni-50%Co)/Al_2O_3.

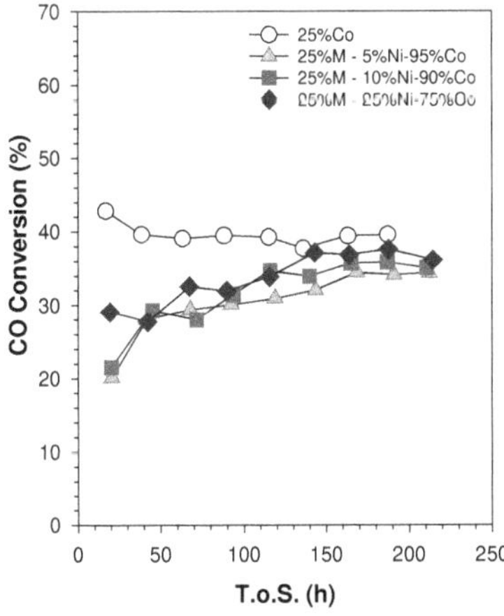

Figure 20. Evolution with T.o.S. of CO conversion for the prepared catalyst (process conditions: T = 220, H_2/CO = 2 mol/mol, P = 20.6 bar, S.V. = 3.4 slph per g_{cat}).

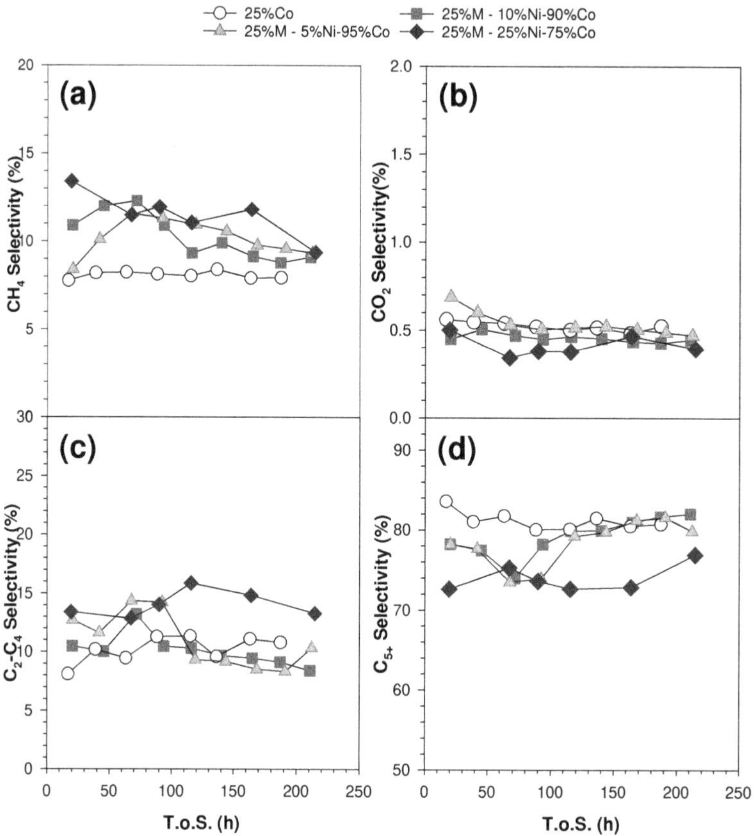

Figure 21. Evolution with T.o.S. of (**a**) CH_4, (**b**) CO_2, (**c**) C_2–C_4, (**d**) and C_{5+} selectivities for the prepared catalyst (process conditions: P = 20.6 bar, H_2/CO = 2 mol/mol, T = 220 °C, S.V. = 3.4 slph per g_{cat}).

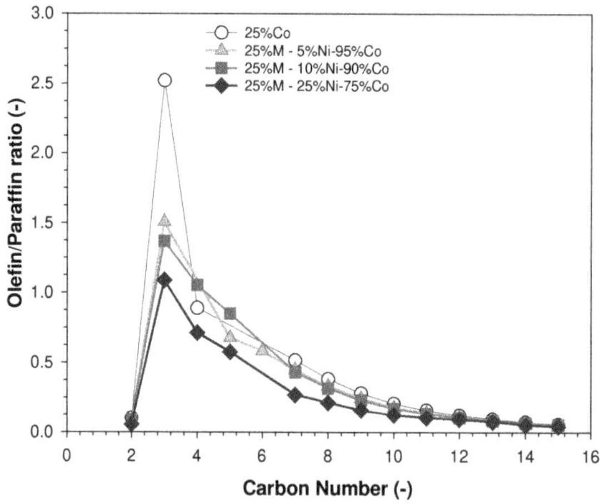

Figure 22. Olefin/paraffin ratio at T.o.S. ≈ 150 h (process conditions: P = 20.6 bar, H_2/CO = 2 mol/mol, T = 220 °C, S.V. = 3.4 slph per g_{cat}).

3. Discussion

The promotion of nickel on a cobalt-based catalyst does not significantly influence the morphological properties, as all the samples have similar surface areas ($\approx$ 92 m^2/g), pore volumes, and pore diameters. The effects of nickel on the structural properties were investigated by XRD and STEM. XRD patterns for all the calcined catalysts showed the typical peaks associated with Co_3O_4, and no diffraction peaks associated with NiO or Ni^0 were observed, suggesting that a mixed metal oxide was formed. Intimate contact between cobalt and nickel was also confirmed by elemental mapping during STEM analysis, as well as by comparing TPR-XANES profiles at nickel and cobalt K-edges. Initially, nickel is associated with Co_3O_4, and then it undergoes a change in the electronic structure to a form of nickel associated with CoO. Further evidence for solid solution formation is that the Ni^{2+} associated with Co_3O_4 (Ni K-edge results) and the Co_3O_4 reduce over a similar temperature range, such that the temperatures at 50% conversion match very well, especially at higher Ni/Co ratios.

TPR-XANES and hydrogen chemisorption/O_2 titration show that nickel has a beneficial effect on cobalt reducibility. The reduction of cobalt oxide is shifted to a lower temperature with increasing Ni/Co ratio, and the percentage of metal reduction is increased. Voss et al. [19] observed an improvement in the reducibility when nickel was added, as the first reduction step shifted to a significantly lower temperature with increasing the nickel loading. The authors proposed a hydrogen spillover effect from the nickel sites to cobalt oxide sites. In the current work, the similarities in temperatures ranges between Co_3O_4 to CoO and Ni^{2+}-Co_3O_4 to Ni^{2+}-CoO, as well as similarities in the second reduction step from CoO to Co^0 and NiO to Ni^0 during TPR-XANES, suggest that in addition to H_2 spillover, a chemical promoting effect exists in leading to the formation of the Co-Ni alloy.

Cobalt-based catalysts are typically characterized by an initial decline and leveling off that occurs within the first few days. Possible mechanisms proposed for deactivation include: reoxidation of small metallic cobalt cluster to inactive CoO_X; sintering; some carbon deposition or solid-state reaction between cobalt and the support [4,25,26]. Thermodynamic calculations show that when the cobalt crystallites have a diameter lower than 4.4 nm, they may be reoxidized in the steam/hydrogen environments of FTS [7,8], but this may occur rapidly before the first measurement is taken. The initial decline and leveling off period for CO conversion typical of cobalt-based catalysts was not observed for the nickel promoted catalysts. Rytter et al. [18] have also observed an activation period in the first 40 h. They proposed a catalyst reconstruction in the first stage of operation where nickel and cobalt partially segregate, thereby exposing the cobalt clusters to FTS reaction. However, additional investigations are needed to speculate the events occurring during this induction time. Interestingly, the performances of nickel-promoted catalysts are stable, despite having lower cobalt loading than 25%Co/Al_2O_3. This suggests that nickel can stabilize cobalt metal nanoparticles. Rytter et al. [18] further proposed that the higher stability for Co-Ni alloys could be due to a suppression in carbon deposition and a suppression of reoxidation phenomena via H_2 spillover.

The addition of nickel changed the product distribution. In particular, the initial methane selectivity increases by increasing the nickel loading, whereas the initial C_{5+} selectivity decreases. Higher CH_4 selectivities were also observed in previous works, where different Ni/Co ratios were studied [10,16,18]. Interestingly, the selectivities of Co/Ni catalysts improve with time on-stream to nearly match those of pure Co catalysts. Furthermore, the olefin content decreases, as does the chain growth probability with increasing Ni/Co ratio. These results are not surprising because of the high hydrogenation capability of nickel. Furthermore, Ishihara at al. [12] studied Co-Ni alloys supported on SiO_2 and found that cobalt electronically interacts with nickel in the outer shell orbitals by creating adsorption sites possessing a new electron density. These new sites have the highest H adsorption strengths. Thus, hydrogen competes more effectively for adsorption sites during co-adsorption of CO and H_2 with Co-Ni alloys relative to the pure cobalt based catalyst.

Finally, it is interesting to compare the activity/$ (Table 4), as nickel and cobalt have different market prices. The price of cobalt is reported to be 2.13 times the price of nickel. Thus, the partial substitution of cobalt with nickel would be an advantage in terms of total catalyst cost. At steady state conditions, the activity/$ for the 25%M—25%Ni-75%Co is best for both the nickel-promoted and unpromoted catalyst; additionally, at that point, the selectivities of the CoNi catalysts nearly match those of the pure Co based catalyst.

Table 4. Activity/$ at steady-state conditions for the tested catalyst.

Sample ID	Activity/$
25%Co	39.5
25%M—5%Ni-95%Co	35.4
25%M—10%Ni-90%Co	36.8
25%M—25%Ni-75%Co	41.9

4. Materials and Methods

4.1. Catalyst Preparation

The conventional slurry impregnation method was used to prepare the catalyst containing 25% metal by weight, with the following Ni/Co atomic ratios: 0/100, 5/95, 10/90, 25/75, and 50/50. The support was Catalox 150 γ-alumina with a surface area of 150 m^2/g. Nickel nitrate and cobalt nitrate (Alfa Aesar) served as the precursors to load the nickel and cobalt together (i.e., in a single solution) onto the γ-Al_2O_3 support. In this method, the ratio of the weight of alumina to the volume of solution used was 1:1, as reported in a Sasol patent [1], such that the loading solution prepared was approximately 2.5 times that of the pore volume. The total metal loading was added by two impregnation steps with 12.5% of metal by weight for each step. Between each step, the catalyst was dried at 60 °C under vacuum in a rotary evaporator; then, the temperature was slowly increased until 100 °C. After the second impregnation, the catalyst was dried and then calcined in flowing air for 4 h at 350 °C.

4.2. Characterization

A Micromeritics 3-Flex system (Norcross, GA, USA) using N_2 (UHP N_2, Airgas, Lexington, KY, USA) for physisorption was used to measure BET surface area and porosity properties. Before testing, the samples were pretreated at 160 °C and 50 mTorr for at least 12 h. The BJH method was employed to calculate the average pore volume and pore diameter.

XRD spectra were collected with a Philips X'Pert diffractometer (Amsterdam, The Netherlands) with monochromatic Cu Kα radiation (λ = 1.54Å). The conditions employed included a scan rate of 0.01° per step, 2θ range of 10–90, and scan time of 4 s per step.

Prior to STEM characterization, the samples were pretreated in hydrogen (American Welding and Gas, Lexington, KY, USA) at 350 °C for 18 h, cooled down to room temperature, and then passivated with a mixture of 1% O_2 in nitrogen (American Welding and Gas, Lexington, KY, USA). STEM analysis was performed with an FEI Talos F200X instrument (Thermo Scientific, Waltham, MA, USA) equipped with BF, DF2, DF4, and HAADF detectors. The imaging was collected with a field emission gun using an accelerating voltage of 200 kV and a high speed Ceta 16M camera; Velox software (Thermo Scientific, Waltham, MA, USA) was used for data processing. The samples were dispersed in ethanol (Alfa Aesar, Haverhill, MA, USA), sonicated, and then dropped onto a carbon-coated copper grid and dried in air.

Hydrogen chemisorption and the ensuing pulse reoxidation were carried out using an Altamira AMI-300 unit (Altamira Instruments, Pittsburgh, PA, USA). The sample was reduced at 350 °C (ramping rate at 2 °C/min) for 10 h in 10 cm^3/min of UHP H_2 (Airgas, San Antonio, TX, USA) blended with 20 cm^3/min of UHP argon (Airgas, San Antonio, TX, USA). Then, the temperature was cooled to 100 °C, and UHP argon (30 cm^3/min) was flowed through the catalytic bed to avoid the adsorption of weakly bound hydrogen. Next, the temperature was increased to 350 °C at 10 °C/min in flowing argon to desorb the chemisorbed hydrogen. The hydrogen peak obtained during the temperature programmed desorption was integrated and the moles of hydrogen evolved was determined by comparing to calibration pulses. Pulses of UHP O_2 (Airgas, San Antonio, TX, USA) were then passed through the reactor to reoxidize the catalyst until saturation was achieved. The percentage of reduction was estimated with two different methods. In the first approach, nickel and cobalt metal were assumed to oxidize to NiO and Co_3O_4, respectively. However, in a second approach, we assumed all Co_3O_4 converted at least to CoO during the reduction, and that a fraction of NiO and CoO converted to Ni^0 and Co^0. Thus, during the reoxidation step, the Ni^0 and Co^0 are first oxidized to NiO and CoO. Then, all CoO oxidizes to Co_3O_4. These two approaches set minimum and maximum limits for the cobalt cluster size when the uncorrected dispersion is modified by considering the percentage of reduction by the metal, as follows:

% Dispersion (uncorrected) = (# metal atoms on the surface)/(# metal atoms in the sample)
% Disp. (corrected) = (# metal atoms on the surface)/((# metal atoms in the sample)(% reduction))

4.3. H_2-TPR XANES-EXAFS

In-situ H_2-TPR XAFS experiments were carried out at the Materials Research Collaborative Access Team (MR-CAT) beamline at the Advanced Photon Source, Argonne National Laboratory. A cryogenically cooled Si (1 1 1) monochromator selected the incident energy and a rhodium-coated mirror rejected higher order harmonics of the fundamental beam energy. The experiment setup was analogous to that outlined by Jacoby [27]. The in-situ TPR of 6 catalysts were performed in stainless-steel multi-sample holder (3.0 mm i.d. channels). Approximately 6 mg of each catalyst was loaded as a self-supporting wafer in each channel. The catalyst was diluted with alumina in a weight ratio of approximately 1:1. The holder was located in the center of a quartz tube and equipped with Kapton windows, a thermocouple, and gas ports. The amount of catalyst loaded was optimized for the Co and Ni K edges, considering the absorption by aluminum of the Al_2O_3. The quartz tube was positioned in a clamshell furnace mounted on the positioning table. Each sample cell was placed relative to the beam, and the position of the table was adjusted to an accuracy of 20 μm (for repeated scans). Once the catalyst positions were fine-tuned, the reactor was purged with He (100 mL/min) for more than 5 min, and then the reactant gas (a mixture H_2/He, 3.5%) was flowed through the samples (100 mL/min). The temperature was increased to 700 °C (ramp of ≈ 1.0 °C/min) and then held for 4 h. The Ni and Co K-edge spectra were collected in transmission mode. The Co metallic foil spectrum was also recorded simultaneously for energy calibration. X-ray absorption spectra for each catalyst were collected from 7500 to 9000 eV.

The WinXAS program was used to analyze the spectra collected during H_2-TPR EXAFS/XANES experiments [28]. Additional details of the EXAFS and XANES analyses for Co K-edge data are reported in our previous work [29]. Ni K-edge data were processed in a similar manner (i.e., same Δk and ΔR in fittings). For qualitative comparisons of EXAFS and XANES results, the references used for Co^0, CoO, and Co_3O_4 were the final spectrum, the point of maximum CoO content, and the initial spectrum of the TPR trajectory of undoped cobalt catalyst (25%Co/Al_2O_3). For Ni^0 and NiO, the references were NiO (Alfa Aesar, Puratronic, 99.998%, Tewksbury, MA, USA) and a Ni^0 foil.

For XANES analyses, linear combination fittings were carried out considering as reference compounds for Co K-edge data, the initial spectrum (a mixture of Co^{3+} and Co^{2+} similar to Co_3O_4), the point of maximum CoO content, and the final spectrum after H_2 TPR (representing Co^0). At the Ni K-edge, the reference compounds were the first spectrum (Ni^{2+} associated with Co_3O_4), the spectrum with highest CoO content, where Ni^{2+} is associated with the CoO, and the final spectrum after H_2 TPR (representing ≈ 100% Ni^0). The data reduction and fitting for EXAFS were performed using the catalysts in their final state following TPR and cooling in flowing H_2 using the WinXAS [28], Atoms [30], FEFFIT [31], and FEFF [31] programs. The k-range chosen for the fittings was 3–10 Å^{-1}. Fitting was confined to the first metallic coordination shell by applying a Hanning window in the Fourier transform magnitude spectra, and carrying out the back-transform to isolate that shell.

4.4. Reaction Testing

Activity tests were performed using a 1 L continuously stirred tank reactor (CSTR) (PPI, Warminster, PA, USA). Additional information on the laboratory-scale rig can be found elsewhere [32]. In a typical test, 9.6 g of catalyst (63 < dp < 125 μm) were loaded into a fixed bed reactor. The catalyst was reduced at 350 °C for 20 h, while being fed a 30 Nl/h H_2/He mixture (1:3 v/v, American Welding and Gas, Lexington, KY, USA) at atmospheric pressure. The reduced catalyst was transferred by pneumatic transfer under the protection of inert gas to a CSTR containing 310 g of melted Polywax 3000 (Baker Petrolite, Houston, TX, USA). In-situ reduction of the transferred catalyst was performed at atmospheric pressure and 230 °C with overnight feeding of 30 Nl/h pure H_2 (American Welding and Gas, Lexington, KY, USA). In this work, the catalytic testing was carried out at the following process conditions: P = 20.2 bar, T = 220 °C, a stirring speed of 750 rpm, and H_2/CO = 2 mol/mol. The unconverted reactants and the products leaving the CSTR were sent to a warm trap, in which the temperature was set at 100 °C, and then to a cold trap maintained at 0 °C. The uncondensed stream was reduced to atmospheric pressure, while the flowrate and the composition were measured by a wet test meter and by an online 3000A micro-GC (Agilent, Santa Clara, CA, USA), respectively. The micro-GC is equipped with four different columns (Plot U, Molecular Sieve, OV-1 and, Alumina) and TCD. The reaction products were collected in three traps maintained at different temperatures: a hot trap (200 °C), a warm trap (100 °C), and a cold trap (0 °C). The products were separated into different fractions (wax, oil, and aqueous) for quantification. The oil (C_4–C_{20}) fraction was analyzed with a 7890 GC (Agilent, Santa Clara, CA, USA) equipped with DB-5 (60 m × 0.32 mm × 0.25μm, Agilent J&W) column and FID, whereas waxes (C_{21}–C_{60}) were analyzed with an HP 6890 GC equipped with ZB-1HT column (30 m × 0.25 mm × 0.10 μm, Zebron) and FID.

5. Conclusions

Bimetallic catalysts with different Ni/Co ratios were prepared by standard aqueous incipient impregnation. N_2 adsorption/desorption results show that the addition of nickel had no effect on the morphological properties, as similar surface areas, pore volumes, and pore diameters were obtained independently from the Ni/Co ratio. XRD patterns of the samples exhibited peaks associated with Co_3O_4, whereas no diffraction peaks associated with Ni were observed. Thus, this suggests a Co–Ni solid oxide solution might be formed. STEM and TPR-XANES show the nickel and cobalt are in intimate contact, strongly suggesting the formation of a Co-Ni alloy from a Co oxide – Ni oxide solid solution. Moreover, TPR-XANES results indicate that nickel promotion improves the cobalt reducibility by systematically shifting the reduction profiles to lower temperatures. The similarities in temperature ranges between Co_3O_4 to CoO and Ni^{2+}-Co_3O_4 to Ni^{2+}-CoO, and the second reduction step to the metallic phase (i.e., CoO to Co^0 and NiO to Ni^0) during TPR-XANES, suggest that not only is H_2 spillover involved, but a chemical effect is likely. This is due to intimate contact of Co and Ni in the solid solution that leads to the formation of the Co-Ni alloy.

The catalyst's performance during FTS, in terms of conversion, selectivity, and stability, was evaluated using a CSTR. Nickel-promoted catalysts have lower initial CO conversion, which progressively increases with T.o.S. until a steady-state value is achieved. This CO conversion trend is inverted compared to the typical induction period observed for Co/Al_2O_3, where the activity progressively declines and levels off because of deactivation phenomena. The stability of the Co–Ni alloy may be due to the stabilization of metallic cobalt nanoparticles by nickel addition, resulting in robust nanoparticles even at lower cobalt content compared to commercial Co loadings. CH_4 selectivity increases by increasing nickel loading because of the higher hydrogenation capability of the Co–Ni alloy. However, the difference in CH_4 selectivity between the Ni-promoted and unpromoted catalyst (in terms of the absolute value) decreases with T.o.S., which is beneficial; in fact, after stabilization, the C_5+ selectivities were quite similar between catalysts prepared with Co–Ni bimetallic mixtures and Co alone. Finally, the steady-state activity/$ for a Ni/Co ratio of 25/75 is slightly higher than that of 25%Co/Al_2O_3.

Author Contributions: Conceptualization, catalyst preparation, catalyst characterization, formal analysis, writing, G.J. Catalyst preparation, catalyst characterization, formal analysis, S.C.K. Catalyst preparation, catalyst characterization, formal analysis, R.G. Catalyst preparation, catalyst characterization, formal analysis. C.D.W. Reaction testing, characterization, formal analysis, conceptualization, writing, M.M. Project administration, resources, C.L.M. Catalyst preparation, supervision, resources, D.C.C. Catalyst characterization, data curation, resources, supervision, A.J.K. Formal analysis W.D.S. All authors have read and agreed to the published version of the manuscript.

Funding: Richard Garcia received funding from a UTSA College of Engineering Scholarship. His work was also supported by the USDA National Institute of Food and Agriculture, Interdisciplinary Hands-on Research Traineeship and Extension Experiential Learning in Bioenergy/Natural Resources/Economics/Rural project, U-GREAT (Undergraduate Research, Education and Training) program (2016-67032-24984). Caleb D. Watson acknowledges support from the Undergraduate NSF Research Program, supported by the National Science Foundation through grant award #1832388.

Acknowledgments: Argonne's research was supported in part by the U.S. Department of Energy (DOE), Office of Fossil Energy, National Energy Technology Laboratory (NETL). Advanced photon source was supported by the U.S. Department of Energy, Office of Science, Office of Basic Energy Sciences, under contract number DE-AC02-06CH11357. MRCAT operations are supported by the Department of Energy and the MRCAT member institutions. CAER research was supported by the Commonwealth of Kentucky. Richard Garcia would like to acknowledge funding from a UTSA College of Engineering Scholarship. His work was also supported by the USDA National Institute of Food and Agriculture, Interdisciplinary Hands-on Research Traineeship, and Extension Experiential Learning in Bioenergy/Natural Resources/Economics/Rural project, U-GREAT (Undergraduate Research, Education and Training) program (2016-67032-24984). Caleb D. Watson would like to acknowledge support from the Undergraduate NSF Research Program, supported by the National Science Foundation through grant award #1832388. Gary Jacobs would like to thank UTSA and the State of Texas for financial support through startup funds.

Conflicts of Interest: The authors declare no conflict of interest.

References

1. Schulz, H. Short history and present trends of Fischer–Tropsch synthesis. *Appl. Catal. A Gen.* **1999**, *186*, 3–12. [CrossRef]
2. Shafer, W.D.; Gnanamani, M.K.; Graham, U.M.; Yang, J.; Masuku, C.M.; Jacobs, G.; Davis, B.H. Fischer–Tropsch: Product selectivity–The fingerprint of synthetic fuels. *Catalysts* **2019**, *9*, 259. [CrossRef]
3. Khodakov, A.Y.; Chu, W.; Fongarland, P. Advances in the development of novel cobalt Fischer−Tropsch catalysts for synthesis of long-chain hydrocarbons and clean fuels. *Chem. Rev.* **2007**, *107*, 1692–1744. [CrossRef] [PubMed]
4. Jahangiri, H.; Bennett, J.; Mahjoubi, P.; Wilson, K.; Gu, S. A review of advanced catalyst development for Fischer Tropsch synthesis of hydrocarbons from biomass derived syn-gas. *Catal. Sci. Technol.* **2014**, *4*, 2210–2229. [CrossRef]
5. Glacier, I. Resource Management Group. Available online: http://www.infomine.com/investment (accessed on 20 December 2019).
6. Iglesia, E. Design, synthesis, and use of cobalt-based Fischer-Tropsch synthesis catalysts. *Appl. Catal. A Gen.* **1997**, *161*, 59–78. [CrossRef]
7. Hou, C.; Xia, G.; Sun, X.; Wu, Y.; Jin, C.; Yan, Z.; Li, M.; Hu, Z.; Nie, H.; Li, D. Thermodynamics of oxidation of an alumina-supported cobalt catalyst by water in F-T synthesis. *Catal. Today* **2016**, *264*, 91–97. [CrossRef]
8. Van Steen, E.; Claeys, M.; Dry, M.E.; Van de Loosdrecht, J.; Viljoen, E.L.; Visagie, J.L. Stability of nanocrystals: Thermodynamic analysis of oxidation and re-reduction of cobalt in water/hydrogen mixtures. *J. Phys. Chem. B* **2005**, *109*, 3575–3577. [CrossRef]
9. Jacobs, G.; Patterson, P.M.; Das, T.K.; Luo, M.; Davis, B.H. Fischer–Tropsch synthesis: Effect of water on Co/Al_2O_3 catalysts and XAFS characterization of reoxidation phenomena. *Appl. Catal. A Gen.* **2004**, *270*, 65–76. [CrossRef]
10. Van Helden, P.; Prinsloo, F.; Van den Berg, J.-A.; Xaba, B.; Erasmus, W.; Claeys, M.; Van de Loosdrecht, J. Cobalt-nickel bimetallic Fischer-Tropsch catalysts: A combined theoretical and experimental approach. *Catal. Today* **2020**, *342*, 88–98. [CrossRef]
11. Shadravan, V.; Bukas, V.J.; Gunasooriya, G.T.K.K.; Waleson, J.; Drewery, M.; Karibika, J.; Jones, J.; Kennedy, E.; Adesina, A.; Nørskov, J.K.; et al. Effect of manganese on the selective catalytic hydrogenation of COx in the presence of light hydrocarbons over Ni/Al_2O_3: An experimental and computational study. *ACS Catal.* **2020**, *10*, 1535–1547. [CrossRef]

12. Ishihara, T.; Horiuchi, N.; Inoue, T.; Eguchi, K.; Takita, Y.; Arai, H. Effect of alloying on CO hydrogenation activity over SiO_2-supported Co-Ni alloy catalysts. *J. Catal.* **1992**, *136*, 232–241. [CrossRef]
13. Ishihara, T.; Iwakuni, H.; Eguchi, K.; Arai, H. Hydrogenation of carbon monoxide over the mixed catalysts composed of cobalt-nickel/manganese oxide-zirconium oxide and zeolite catalysts. *Appl. Catal.* **1991**, *75*, 225–235. [CrossRef]
14. Ishihara, T.; Eguchi, K.; Arai, H. Hydrogenation of carbon monoxide over SiO_2-supported Fe-Co, Co-Ni and Ni-Fe bimetallic catalysts. *Appl. Catal.* **1987**, *30*, 225–238. [CrossRef]
15. Sun, Y.; Wei, J.; Zhang, J.P.; Yang, G. Optimization using response surface methodology and kinetic study of Fischer–Tropsch synthesis using SiO_2 supported bimetallic Co–Ni catalyst. *J. Nat. Gas Sci. Eng.* **2016**, *28*, 173–183. [CrossRef]
16. Shimura, K.; Miyazawa, T.; Hanaoka, T.; Hirata, S. Fischer–Tropsch synthesis over alumina supported bimetallic Co–Ni catalyst: Effect of impregnation sequence and solution. *J. Mol. Catal. A Chem.* **2015**, *407*, 15–24. [CrossRef]
17. Eshraghi, A.; Mirzaei, A.A.; Rahimi, R.; Atashi, H. Effect of Ni–Co morphology on kinetics for Fischer–Tropsch reaction in a fixed-bed reactor. *J. Taiwan Inst. Chem. Eng.* **2019**, *105*, 104–114. [CrossRef]
18. Rytter, E.; Skagseth, T.H.; Eri, S.; Sjåstad, A.O. Cobalt Fischer−Tropsch catalysts using nickel promoter as a rhenium substitute to suppress deactivation. *Ind. Eng. Chem. Res.* **2010**, *49*, 4140–4148. [CrossRef]
19. Voss, G.J.B.; Fløystad, J.B.; Voronov, A.; Rønning, M. The state of nickel as promotor in cobalt Fischer–Tropsch synthesis catalysts. *Top. Catal.* **2015**, *58*, 896–904. [CrossRef]
20. López-Tinoco, J.; Mendoza-Cruz, R.; Bazán-Díaz, L.; Karuturi, S.C.; Martinelli, M.; Cronauer, D.C.; Kropf, A.J.; Marshall, C.L.; Jacobs, G. The preparation and characterization of Co–Ni nanoparticles and the testing of a heterogenized Co–Ni/alumina catalyst for CO hydrogenation. *Catalysts* **2019**, *10*, 18. [CrossRef]
21. Nikparsa, P.; Mirzaei, A.A.; Rauch, R. Modification of Co/Al_2O_3 Fischer–Tropsch nanocatalysts by adding Ni: A kinetic approach. *Int. J. Chem. Kinet.* **2016**, *48*, 131–143. [CrossRef]
22. Yu, H.; Zhao, A.; Zhang, H.; Ying, W.; Fang, D. Bimetallic catalyst of Co and Ni for Fischer-Tropsch synthesis supported on alumina, energy sources. *Part A Recovery Util. Environ. Eff.* **2015**, *37*, 47–54.
23. Jacobs, G.; Ma, W.; Gao, P.; Todic, B.; Bhatelia, T.; Bukur, D.B.; Davis, B.H. The application of synchrotron methods in characterizing iron and cobalt Fischer–Tropsch synthesis catalysts. *Catal. Today* **2013**, *214*, 100–139. [CrossRef]
24. Van Der Laan, G.P.; Beenackers, A.A.C.M. Kinetics and selectivity of the Fischer–Tropsch synthesis: A literature review. *Catal. Rev.* **1999**, *41*, 255–318. [CrossRef]
25. Tsakoumis, N.E.; Rønning, M.; Borg, Ø.; Rytter, E.; Holmen, A. Deactivation of cobalt based Fischer–Tropsch catalysts: A review. *Catal. Today* **2010**, *154*, 162–182. [CrossRef]
26. Jacobs, G.; Patterson, P.M.; Zhang, Y.; Das, T.; Li, J.; Davis, B.H. Fischer–Tropsch synthesis: Deactivation of noble metal-promoted Co/Al_2O_3 catalysts. *Appl. Catal. A Gen.* **2002**, *233*, 215–226. [CrossRef]
27. Jacoby, M. X-ray absorption spectroscopy. *Chem. Eng. News Arch.* **2001**, *79*, 33–38. [CrossRef]
28. Ressler, T. WinXAS: A program for X-ray absorption spectroscopy data analysis under MS-Windows. *J. Synchrotron Radiat.* **1998**, *5*, 118–122. [CrossRef]
29. Jacobs, G.; Ji, Y.; Davis, B.H.; Cronauer, D.; Kropf, A.J.; Marshall, C.L. Fischer–Tropsch synthesis: Temperature programmed EXAFS/XANES investigation of the influence of support type, cobalt loading, and noble metal promoter addition to the reduction behavior of cobalt oxide particles. *Appl. Catal. A Gen.* **2007**, *333*, 177–191. [CrossRef]
30. Ravel, B. ATOMS: Crystallography for the X-ray absorption spectroscopist. *J. Synchrotron Radiat.* **2001**, *8*, 314–316. [CrossRef]
31. Newville, M.; Ravel, B.; Haskel, D.; Rehr, J.J.; Stern, E.A.; Yacoby, Y. Analysis of multiple-scattering XAFS data using theoretical standards. *Phys. B Condens. Matter* **1995**, *208–209*, 154–156. [CrossRef]
32. Jacobs, G.; Ribeiro, M.C.; Ma, W.; Ji, Y.; Khalid, S.; Sumodjo, P.T.A.; Davis, B.H. Group 11 (Cu, Ag, Au) promotion of 15%Co/Al_2O_3 Fischer–Tropsch synthesis catalysts. *Appl. Catal. A Gen.* **2009**, *361*, 137–151. [CrossRef]

© 2020 by the authors. Licensee MDPI, Basel, Switzerland. This article is an open access article distributed under the terms and conditions of the Creative Commons Attribution (CC BY) license (http://creativecommons.org/licenses/by/4.0/).

 *catalysts* 

Article

Manganese and Cobalt Doped Hierarchical Mesoporous Halloysite-Based Catalysts for Selective Oxidation of *p*-Xylene to Terephthalic Acid

Eduard Karakhanov [1], Anton Maximov [1,2], Anna Zolotukhina [1,2], Vladimir Vinokurov [3], Evgenii Ivanov [3] and Aleksandr Glotov [3,*]

1 Department of Petroleum Chemistry and Organic Catalysis, Moscow State University, 119991 Moscow, Russia; kar@petrol.chem.msu.ru (E.K.); max@ips.ac.ru (A.M.); anisole@yandex.ru (A.Z.)
2 A.V. Topchiev Institute of Petrochemical Synthesis, Russian Academy of Sciences, 119991 Moscow, Russia
3 Department of Physical and Colloid Chemistry, Gubkin Russian State University of Oil and Gas, 119991 Moscow, Russia; vinok_ac@mail.ru (V.V.); ivanov166@list.ru (E.I.)
* Correspondence: glotov.a@gubkin.ru

Received: 18 November 2019; Accepted: 14 December 2019; Published: 18 December 2019 

Abstract: Bimetallic MnCo catalyst, supported on the mesoporous hierarchical MCM-41/halloysite nanotube composite, was synthesized for the first time and proved its efficacy in the selective oxidation of *p*-xylene to terephthalic acid under conditions of the AMOCO process. Quantitative yields of terephthalic acid were achieved within 3 h at 200–250 °C, 20 atm. of O_2 and at a substrate to the Mn + Co ratio of 4–4.5 times higher than for traditional homogeneous system. The influence of temperature, oxygen, pressure and KBr addition on the catalyst activity was investigated, and the mechanism for the oxidation of *p*-toluic acid to terephthalic acid, excluding undesirable 4-carboxybenzaldehyde, was proposed.

Keywords: halloysite; hierarchical materials; *p*-xylene oxidation; terephthalic acid

1. Introduction

Terephthalic acid (TPA) and its dimethyl ester are important monomers in thermoresistant and mechanically stable polymer production, such as polyethylene terephthalate (PET) and poly-paraphenylene terephthalamide (Kevlar) [1,2]. The world production of terephthalic acid has exceeded 5 Mt/yr. is still growing [1,2]. Currently up to 70% of terephthalic acid is produced by direct oxidation of *p*-xylene through the AMOCO/Mid-Century process being developed in the 1960s [1–6].

p-xylene in the AMOCO process is subjected to oxidation by molecular oxygen or air in the presence of homogeneous Mn and Co catalysts with KBr as a promotor and free-radical source in acetic acid medium under the severe conditions (170–220 °C, 15–30 atm of O_2 or 50–70 atm of air), giving terephthalic acid yield of 90–97% within 4–12 h (Scheme 1) [7–10].

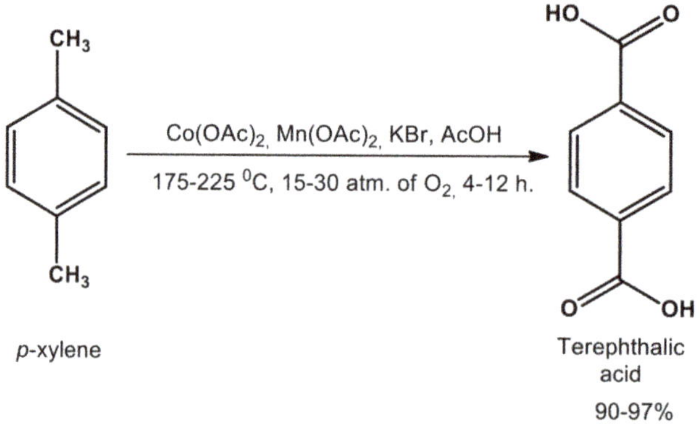

CH_3 … CH_3 — *p*-xylene

$Co(OAc)_2$, $Mn(OAc)_2$, KBr, AcOH

175-225 ^0C, 15-30 atm. of O_2, 4-12 h.

Terephthalic acid 90-97%

Scheme 1. Oxidation of *p*-xylene to terephthalic acid in the AMOCO process [1,2,7,8].

In spite of the obvious advantage of the AMOCO process, such as quantitative yield of high purity terephthalic acid, acetic acid medium with bromides cause corrosion that makes it necessary to use expensive titanium reactors [11–13]. It challenges the development of new environmentally friendly, safer and less corrosive reaction media, promoters and additives [7]. Catalyst heterogenization is considered as one of the possible ways for improving the AMOCO process.

Thus, it was earlier demonstrated, that the use of CoO and Co_3O_4 as catalysts together in the presence of MnO, NiO or CeO co-catalysts and *p*-toluic acid as a promotor allowed for oxidation without KBr [14]. Nonetheless, this system appeared to be much less effective in comparison with the conventional homogeneous system, including $Mn(OAc)_2$, $Co(OAc)_2$ and KBr, and the yield of terephthalic acid did not exceed 65–70% within 8 h [14].

In the presence of MCM-41 doped with Fe and Cu, terephthalic acid underwent further oxidation to 2,5-dihydroxy-1,4-terephthalic and p-benzoquinone-2,5-dicarboxylic acids even under much milder conditions, than in the AMOCO process (H_2O_2 as an oxidant, 80 °C, AcOH, CH_3CN, 5 h) [15]. The highest selectivity to terephthalic acid did not exceed 45% at a *p*-xylene conversion of 10% in the neat acetonitrile. It was found out, that iron additive increased the selectivity to terephthalic acid, whereas copper addition, vice versa, favored the further oxidation of TPA to 2,5-dihydroxy-1,4-terephthalic and p-benzoquinone-2,5-dicarboxylic acids [15].

High yields of terephthalic acid (99% within 2 h) were obtained in the presence of polynuclear μ_3-oxo-bound complexes of Co and Mn, encapsulated in the cavities of Y zeolite, under the conditions similar to AMOCO process (KBr, 200 °C, 610 atm of the air) [16]. The said catalyst appeared as highly resistant to the metal leaching due to close sizes of polyoxo metal clusters and zeolite cavities [8,16].

In this connection, halloysite-based materials appeared to be promising as heterogeneous catalysts for *p*-xylene oxidation under the typical conditions. Halloysite is a natural clay with the rolled tubular structure, appearing as a multiwall nanotube (halloysite nanotube, HNT) with a length of 0.5–1.5 mm, an outer diameter of 50–60 nm and an inner cavity diameter of 10–15 nm (Figure 1) [17,18]. Halloysite clays were successfully applied as carriers for the tubular Ru nanocatalysts, revealing high activity in the hydrogenations of aromatic compounds and phenols [19–23].

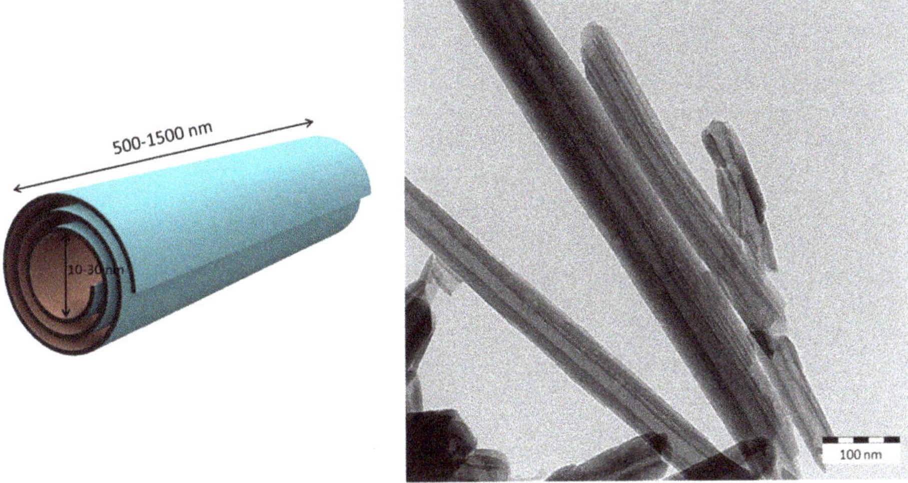

Figure 1. Schematic visualization (**left**) and TEM image (**right**) of halloysite clay.

Grafting of the ordered mesoporous materials, such as MCM-41 or SBA-15, onto halloysite template allows us to obtain new hierarchical systems with stronger mechanical properties and surface area up to 650 m^2/g (Figure 2) [21,24,25]. La-doped MCM-41/HNT composite revealed high efficacy as a sulfur-reducing additive for FCC (fluid catalytic cracking) catalyst, resulting in decrease of sulfur content by 25% and in the yield of gasoline fraction of about 45% [24,26,27]. Modified with CaO and MgO, MCM-41/HNT and SBA-15/HNT composites demonstrated high activity in the cracking of sulfones, formed after the oxidative desulfurization of diesel fraction, decreasing sulfur content from 450 up to 100 ppm [28]. The catalysts said were recycled several times without significant loss of activity, and with high resistance to metal leaching and structure maintenance under the reaction conditions [28].

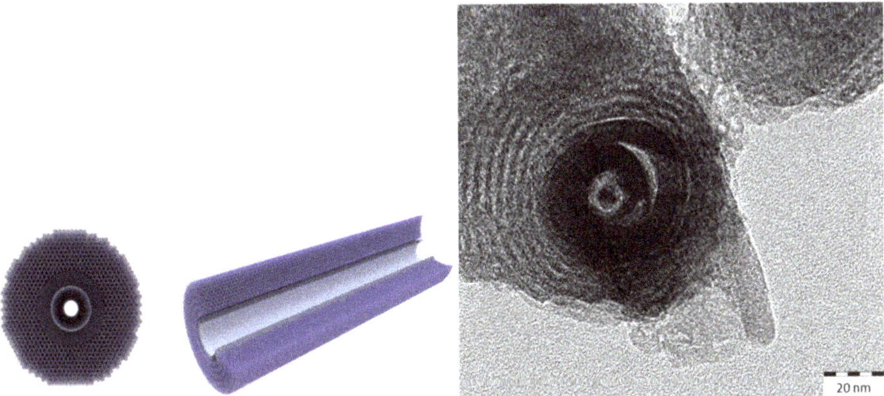

Figure 2. Schematic visualization (**left**) and TEM image (**right**) of MCM-41/HNT composite.

In this work, we present for the first-time synthesis of new heterogeneous bimetallic MnCo catalyst, based on mesoporous hierarchical MCM-41/HNT composite that can be successfully applied for quantitative oxidation of *p*-xylene to terephthalic acid in the AMOCO process, proving its efficiency, that was 4–4.5 times higher, as compared with a traditional homogeneous system.

2. Results and Discussion

2.1. The Synthesis and Characterization of Hierarchical Mesoporous MCM-41/HNT Composite, Doped with Mn and Co

Bimetallic MnCo catalyst, based on MCM-41/HNT composite was synthesized by the wetness impregnation method. Herein Mn^{2+} and Co^{2+} were deposited from the water solution of $Mn(OAc)_2$ and $Co(OAc)_2$ in molar ratio of 1:10 (Scheme 2) [16]. $Mn(OAc)_2$ and $Co(OAc)_2$ tetrahydrates were chosen as sources for Mn^{2+} and Co^{2+} respectively to avoid the influence of other counter-anions on the adsorption and oxidation processes, and because of acetic acid, used as a solvent in the conventional oxidation process [2,8].

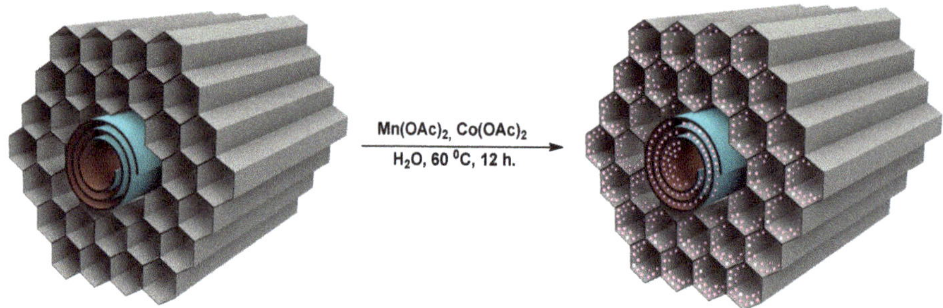

Scheme 2. $Mn(OAc)_2$ and $Co(OAc)_2$ deposition onto MCM-41/HNT composite.

The material obtained was characterized by atomic emission spectroscopy with inductively coupled plasma (ICP-AES), transmission electron microscopy (TEM) and X-ray photoelectron spectroscopy (XPS). The physical and chemical properties are listed in Table 1.

Table 1. Physical and chemical properties of the synthesized $Mn^{II}{}_1Co^{II}{}_{10}$@MCM-41/HNT catalyst.

Mn (wt. %)	Co (wt. %)	Surface Concentration by XPS, at. %							Mn Valency States at Mn $2p_{3/2}$ Line, at. %		Co Valency States at Co $2p_{3/2}$ Line, at. %	
		Mn	Co	Si	Al	C	N	O	MnO (eV)	$[MnO_4]$ Bound (eV)	CoO (eV)	$[CoO_6]$ Bound (eV)
0.15	1.29	<0.1	0.7	25.9	2.5	5.8	0.6	64.2	79.8 (642.5)	20.2 (646.9)	15.3 (780.0)	84.7 (782.5)

As seen from Table 1, weight content of Mn and Co in the sample reached 0.15% and 1.29% respectively, which was approximately three times less than corresponding theoretical values. Herein Co/Mn ratio appeared as 8.6:1, which was in accordance with the literature data [29].

According to XPS data (Table 1, Figure 3), both Mn and Co were presented in the forms of simple oxides MnO [30–32] and CoO [33–35] and bound complexes $[MnO_4]$ [36,37] and $[CoO_6]$ [32–34,38–41], arising from initial $Mn(OAc)_2$ and $Co(OAc)_2$ tetrahydrates as well as from the aluminosilicate. Herein MnO form strongly predominated over $[MnO_4]$ bound form for Mn, whereas for Co $[CoO_6]$ form, vice versa, predominated over CoO, that might be due to the much stronger oxygen affinity for Co in comparison with Mn [42,43]. Taking in account the presence of nitrogen in the XPS spectra (Table 1), carbon in the sample, found out in $–CH_2CH_2–$, $–CH_2CH_2N–$ and $H_3CC(=O)–$ forms [44,45], may be related not only to adsorbed acetate anions, but also to cetyl trimethyl ammonium bromide template, partly remained in the MCM-41/HNT composite after calcination.

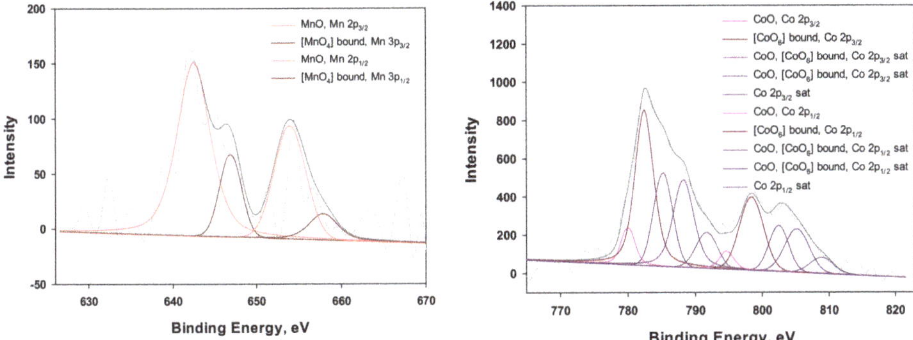

Figure 3. XPS spectra of Mn (**left**) and Co (**right**) in $Mn^{II}{}_1Co^{II}{}_{10}$@MCM-41/HNT composite at 2p line.

As TEM analysis showed (Figure 4), deposition of manganese and cobalt acetates did not destruct the MCM-41/HNT matrix. The MCM-41 framework observed consisted of regular hexagonal pores of 2.9 nm in diameter (Figure 4B) and inner located halloysite nanotubes with the inner diameter of 12–15 nm and outer diameter of 31–35 nm, and interplanar space of 3.2 nm (Figure 4A).

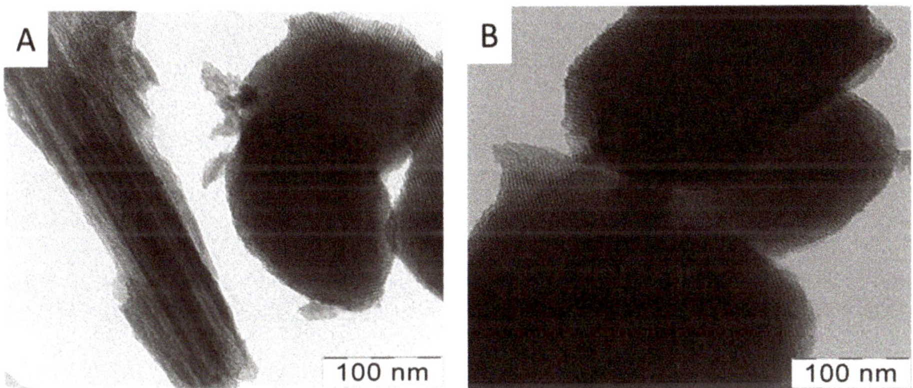

Figure 4. TEM images of MCM-41/HNT composite (A—HNT templated MCM-41, B—MCM-41) after deposition of $Mn(OAc)_2$ and $Co(OAc)_2$.

2.2. Oxidation of p-Xylene in the Presence of Hierarchical Mesoporous MCM-41/HNT Composite Doped with Mn and Co

The catalyst synthesized was tested in the liquid-phase oxidation of *p*-xylene and compared with traditional homogeneous system $Mn(OAc)_2/Co(OAc)_2$/KBr. It should be noted, that *p*-xylene oxidation is a multi-stage radical process (Scheme 3) [2,7,8,46,47]. Its kinetics is strongly affected by the reaction conditions, such as temperature, oxygen pressure, substrate concentration, substrate: Co:Mn ratios, KBr loading, etc. [2,7,8]. Herein KBr is required as a free radical source, and $Mn(OAc)_2$ acts as promotor at the first stage of reaction—activation of the methyl group in the *p*-xylene molecule [2,10]. As a rule, *p*-xylene conversion to *p*-toluic acid proceeds fast, whereas oxidation thereof to terephthalic acid is a slow highly activated process [2,8]. The use of acetic acid as a solvent promotes dissolution of both KBr and *p*-xylene, the ion-radical oxidation process, and deposition of terephthalic acid due to its poor solubility. The TPA product is easily isolated from the reaction medium as white needle-like glittering crystals [2].

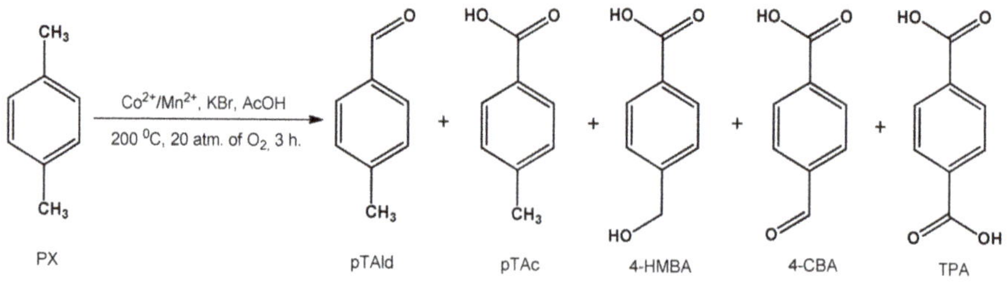

PX $\xrightarrow[\text{200 °C, 20 atm. of } O_2\text{, 3 h.}]{Co^{2+}/Mn^{2+}\text{, KBr, AcOH}}$ pTAld + pTAc + 4-HMBA + 4-CBA + TPA

$4Co^{2+} + O_2 + 4H^+ \rightarrow 4Co^{3+} + 2H_2O$.	$4Co^{2+} + O_2 + 4H^+ \rightarrow 4Co^{3+} + 2H_2O$
$Co^{3+} + Mn^{2+} \rightarrow Co^{2+} + Mn^{3+}$	$Co^{3+} + Mn^{2+} \rightarrow Co^{2+} + Mn^{3+}$
$Mn^{3+} + Br^- \rightarrow Mn^{2+} + Br^\cdot$	$Mn^{3+} + Br^- \rightarrow Mn^{2+} + Br^\cdot$
$ArCH_3 + Br^\cdot \rightarrow ArCH_2^\cdot + H^+ + Br^-$	$ArCH_3 + Br^\cdot \rightarrow ArCH_2^\cdot + H^+ + Br^-$
$ArCH_2^\cdot + O_2 \rightarrow ArCH_2OO^\cdot$	$ArCH_2^\cdot + O_2 \rightarrow ArCH_2OO^\cdot$
$ArCH_2OO^\cdot + Co^{2+} \rightarrow ArC(=O)H + Co^{3+} + OH^-$	$ArCH_2OO^\cdot + RH \rightarrow ArCH_2OOH + R^\cdot$
$ArC(=O)H + Co^{3+} \rightarrow ArC(=O)^\cdot + Co^{2+} + H^+$	$ArCH_2OOH + \rightarrow ArCH_2O^\cdot + OH^\cdot$
$ArC(=O)^\cdot + O_2 \rightarrow ArC(=O)OO^\cdot$	$ArCH_2O^\cdot + RH \rightarrow ArCH_2OH + R^\cdot$
$ArC(=O)OO^\cdot + Ar'CH_3 \rightarrow ArC(=O)OOH + Ar'CH_2^\cdot$	$ArCH_2OH + O_2 \rightarrow ArCH(^\cdot)OH + {}^\cdot OOH$
$ArC(=O)OOH + Co^{2+} \rightarrow ArC(=O)O^\cdot + Co^{3+} + OH^-$	$ArCH_2OH + Br^\cdot \rightarrow ArCH(^\cdot)OH + H^+ + Br^-$
$ArC(=O)O^\cdot + Ar'CH_3 \rightarrow ArC(=O)OH + Ar'CH_2^\cdot$	$ArCH(^\cdot)OH + O_2 \rightarrow ArC(OO^\cdot)OH$
$Ar(=O)O^\cdot + Ar'C(=O)H \rightarrow ArC(=O)OH + Ar'C(=O)^\cdot$	$ArC(OO^\cdot)OH + Co^{2+} \rightarrow ArC(=O)OH + Co^{3+} + OH^-$

Scheme 3. Radical involved mechanism of *p*-xylene oxidation in the presence of Mn^{2+}/Co^{2+}/KBr/AcOH system [2,7,8].

The best conditions for *p*-xylene oxidation, found out in academic studies and widely used in industry (AMOCO process) are as follows: temperature of 180–225 °C, oxygen pressure of 15–30 atm, *p*-xylene to acetic acid molar ratio of 0.08–0.16, and molar ratios Mn/Co and Co:Br being 1:8–10 and 3–4 respectively [2,7,8,48–50]. We have used a temperature of 200 °C, oxygen pressure of 20 atm and molar ratios *p*-xylene/acetic acid and Mn/Co of 0.08 and 1:10 respectively, both in homogeneous and heterogeneous processes. Herein the reaction turnover frequency (TOF) was calculated as the amount of substrate reacted (ν_{substr}) per mole of metal ($\nu_{(Mn + Co)}$) per unit of time with account of each product yield and the number of oxygen atoms added, according to formula:

$$\mathrm{TOF}(\mathrm{O_2}) = \frac{\nu_{substr} * (\omega_{pTAld}*1 + \omega_{\mathrm{pTAc}}*2 + \omega_{\mathrm{4-CBA}}*3 + \omega_{\mathrm{TPA}}*4 + \omega_{\mathrm{N/I}}*3)}{\nu_{(Mn+Co)} * t},$$

where ω is the yield of the certain product, expressed in the unit fractions, N/I is a not identified product (presumably 4-hydroxymethylbenzoic acid) and t is the minimal reaction time, for which the reaction progress is measured. The results obtained are listed in Table 2.

Table 2. Comparison of homogeneous system $Mn(OAc)_2/Co(OAc)_2$ and heterogeneous $Mn^{II}{}_1Co^{II}{}_{10}$@MCM-41/HNT catalyst in the liquid-phase oxidation of *p*-xylene [1].

Entry	Catalyst	Conv., %	Product Yields, % mol.					TOF, h^{-1}
			pTald	pTac	4-HMBA	4-CBA	TPA	
1	$Mn(OAc)_2/Co(OAc)_2$	98.0	0.15	0.95	1.2	0.5	95.2	37.5
2	$Mn^{II}{}_1Co^{II}{}_{10}$@MCM-41/HNT	99.0	0.2	1.0	4.0	-	93.8	142.5

[1] Reaction conditions are: 0.5 mL (4.055 mmol) of *p*-xylene, 5 mL of AcOH, 4.5 mg (0.038 mmol) of KBr, 200 °C, 20 atm of O_2, 3 h–for both homogeneous and heterogeneous processes; 32 mg (0.127 mmol) of $Co(OAc)_2$ and 3.1 mg (0.13 mmol) of $Mn(OAc)_2$ for homogeneous system; 150 mg of bimetallic catalyst for the heterogeneous system.

As one can see, both the homogeneous $Mn(OAc)_2/Co(OAc)_2$ system and heterogeneous $Mn^{II}{}_1Co^{II}{}_{10}$@MCM-41/Hall catalyst give a quantitative conversion and similar product distribution with a TPA yield of 94–95% within 3 h (Table 2). Similar results were obtained earlier for μ-oxo-bridged complexes of Mn and Co, encapsulated into the zeolite Y pores [16]. It should be noted, that near to the quantitative yield of terephthalic acid in the presence of heterogeneous hierarchical catalyst $Mn^{II}{}_1Co^{II}{}_{10}$@MCM-41/HNT was achieved at much higher substrate to the Mn + Co ratio, giving rise to the higher TOF value (Table 2). It proves the efficacy of the new heterogeneous catalyst developed, comprising of the MCM-41/HNT composite, for the AMOCO process. On the other hand, due to the same loading of KBr for both homogeneous and heterogeneous systems, Co:Br in the latter appeared as high as 1:1, resulting in the increased concentration of Br· radicals and, as a consequence, the easier activation of methyl groups in *p*-xylene and *p*-toluic acid molecules.

Temperature, oxygen, pressure and the presence of KBr were found to be crucial factors for the effective *p*-xylene oxidation process, which was in accordance with the literature data [29]. As seen from Table 3, decrease in oxygen pressure down to 5 atm resulted in corresponding decrease in conversion to 88–89%, and *p*-toluic acid appeared as the major reaction product with the yield of 50–63% for both homogeneous and heterogeneous catalysts. Moreover, lower metal loading in heterogeneous system gave rise to a lower rate of further oxidation of *p*-toluic acid, characterized by the higher activation energy [8], resulting in the extremely low yield of terephthalic acid, about 0.5% (Table 3, Entry 2).

Table 3. The influence of the oxygen pressure and KBr presence on the effectiveness of the *p*-xylene oxidation [1].

Entry	Catalyst	$P(O_2)$, atm	KBr	Conv., %	Product Yields, % mol.						TOF, h^{-1}
					pTald	pTac	TPAld [2]	4-HMBA	4-CBA	TPA	
1	$Mn(OAc)_2/Co(OAc)_2$	5	Yes	89.1	11.7	49.8	x	12.6	11.8	3.2	19.0
2	$Mn^{II}{}_1Co^{II}{}_{10}$@MCM-41/HNT	5	Yes	87.9	14.1	62.8	0.25	6.2	4.1	0.4	63.2
3	$Mn^{II}{}_1Co^{II}{}_{10}$@MCM-41/HNT	20	No	2.2	1.1	0.4	-	0.1	0.1	0.5	1.6

[1] Reaction conditions are: 0.5 mL (4.055 mmol) of *p*-xylene, 5 mL of AcOH, 4.5 mg (0.038 mmol) of KBr, 200 °C, 3 h—for both homogeneous and heterogeneous processes; 150 mg of bimetallic catalyst for heterogeneous system; 32 mg (0.127 mmol) of $Co(OAc)_2$ and 3.1 mg (0.13 mmol) of $Mn(OAc)_2$ for homogeneous system. [2] TPAld is terephthalic aldehyde.

The removal of KBr from the reaction medium led to abrupt downfall in conversion for heterogeneous $Mn^{II}{}_1Co^{II}{}_{10}$@MCM-41/HNT catalyst even at high oxygen pressure, when all other conditions are equal (Table 3, Entry 3). *p*-Toluic aldehyde was found to be the major reaction product with the yield of 1% only. The results obtained seemed to be connected not only with the absence of the source of Br· radicals, initiating the oxidation stepwise process, but also with the extremely low concentration of Mn^{3+} ions in the system, being also responsible for methyl group activation in the substrate. For the successful oxidation of *p*-xylene, Pd-containing systems were suggested [51–53] or compounds, such as *N*-hydroxyphthalimide, being able to generate stable free radicals [8,54,55]. In these cases, however, a very long time (12–48 h) is needed to obtain the yield of terephthalic acid, exceeding 70%.

The influence of temperature on the rate of the *p*-xylene oxidation and product distribution in the presence of $Mn^{II}{}_1Co^{II}{}_{10}$@MCM-41/HNT heterogeneous catalyst is presented in Table 4. As one can see, the quantitative yield of terephthalic acid may be obtained in 5 h at 200 °C and in 3 h at 250 °C, and the TOF values being 401.7 and 424.8 h^{-1} respectively (Table 4, Entries 6 and 8–9). Hence, an increase in temperature from 200 to 250 °C results in a slight rise of the catalyst activity and, therefore, the oxidation rate only.

Table 4. The influence of the temperature on the effectiveness of the *p*-xylene oxidation in the presence of $Mn^{II}{}_1Co^{II}{}_{10}$@MCM-41/HNT catalyst [1].

Entry.	T, °C	t, h	Conv., %	Product Yields, % mol.					TOF, h^{-1}
				pTald	pTac	4-HMBA	4-CBA	TPA	
1	150	5	36.8	28.8	5.2	1.9	0.1	0.8	15.7
2	150	3	25.9	21.1	2.5	1.8	-	0.5	
3	150	1	10.2	8.0	0.5	1.5	-	0.2	
4	200	5	99.8	0.1	0.2	2.3	-	97.2	401.7
5	200	3	99.0	0.2	1.0	4.0	-	93.8	
6	200	1	98.8	1.9	11.7	0.2	-	85.0	
7	250	5	100	-	-	0.5	-	99.5	424.8
8	250	3	99.9	-	0.5	1.8	-	97.6	
9	250	1	99.5	-	4.5	2.1	-	92.9	

[1] Reaction conditions are: 150 mg of bimetallic MnCo heterogeneous MCM-41/HNT-based catalyst, 0.5 mL (4.055 mmol) of *p*-xylene, 5 mL of AcOH, 4.5 mg (0.038 mmol) of KBr and 20 atm. of O_2.

Vice versa, decrease in temperature down to 150 °C led to an abrupt downfall in activity, in accordance with literature data [29,50]. Herein the *p*-xylene conversion did not exceed 40% even after 5 h, with *p*-toluic aldehyde being obtained as the major reaction product (Table 4, Entry 3).

As seen from Tables 2–4, the yield of 4-carboxybenzaldehyde in the presence of heterogeneous $Mn^{II}{}_1Co^{II}{}_{10}$@MCM-41/Hall catalyst (under 20 atm of O_2) did not exceed 0.5%, and 4-hydroxymethylbenzoic acid being formed as the main by-product with the yield up to 5%. At low oxygen pressures the yield of 4-carboxybenzaldehyde increased and reached 12% for homogeneous $Mn(OAc)_2/Co(OAc)_2$ system and 4% for heterogeneous $Mn^{II}{}_1Co^{II}{}_{10}$@MCM-41/HNT catalyst (Table 3, Entries 1–2). It should be noted, that 4-carboxybenzaldehyde is the undesirable product of *p*-xylene oxidation: it is slowly oxidized to terephthalic acid and cocrystallizes with it at separation due to structural similarity [2,7,8].

One may assume, that, when $Mn^{II}{}_1Co^{II}{}_{10}$@MCM-41/Hall is used as the catalyst, the oxidation passed through the intermediate alcohol formation (Scheme 3, right). Due to the low metal content and/or specific microenvironment of active sites in the $Mn^{II}{}_1Co^{II}{}_{10}$@MCM-41/HNT catalyst, caused by the interaction of Mn (II) and Co (II) with halloysite and/or MCM-41 matrix, $ArCH_2OO\cdot$ radical apparently interacted not with Co^{2+}, resulting in aldehyde formation, but with each possible reactant molecule, resulting in the alcohol formation. Noticeably, the oxidation of *p*-xylene to *p*-toluic acid mainly proceeded through the formation of *p*-toluic aldehyde, and only traces of *p*-toluic alcohol were observed at low conversions of *p*-xylene. Hence, the "alcohol pathway" was presumably the characteristic of the oxidation *p*-toluic acid. Thus obtained 4-hydroxymethylbenzoic acid underwent further Co (II) catalyzed oxidation to terephthalic acid, initiated by O_2 or Br· as radicals.

We think the explanation above is consistent with the influence of oxygen pressure and KBr observed, as well as with a high reaction rate even at very low metal loading, for the selective oxidation of *p*-xylene to terephthalic acid in the presence of heterogeneous $Mn^{II}{}_1Co^{II}{}_{10}$@MCM-41/HNT catalyst. Therefore, MCM-41/HNT composite, doped with Mn and Co, may be considered as a prospective catalyst for the AMOCO process, far superior to the original homogeneous $Mn(OAc)_2/Co(OAc)_2$ system.

Unfortunately, metal leaching took place during the reaction. The quantity of leached metals depended on the reaction temperature and time: a higher temperature and longer reaction time resulted in less metal maintained in the catalyst (Table 5). Herein Co leached the first followed by Mn. When carrying out the reaction for the longest time (5 h), the concentrations of both Mn and Co dropped approximately by 13–15 times (Table 5, Entry 5). XPS analysis revealed no Mn or Co on the surface of the samples recycled.

Table 5. Comparative Mn and Co weight content in the recycled samples of $Mn^{II}{}_1Co^{II}{}_{10}$@MCM-41/HNT catalyst depending on reaction conditions.

Entry	Mn (wt. %)	Co (wt. %)	Reaction Conditions
1	0.15	1.29	Initial catalyst
2	0.15	0.22	150 °C, 20 atm. of O_2, 3 h
3	0.03	0.13	200 °C, 20 atm. of O_2, 3 h
4	0.04	0.09	250 °C, 20 atm. of O_2, 3 h
5	0.01	0.10	200 °C, 20 atm. of O_2, 5 h

We suppose, the metal leaching observed occurs due to dissolution of the adsorbed Mn and Co species by acetic and hydrobromic acids. The last one is resulted from the interaction of Br· radicals with substrate or intermediate compounds, bearing H-atoms at side carbons (Scheme 3). Nonetheless it should also be noted, that catalyst, based on MCM-41/HNT composite possesses very low packed density and, simultaneously, high apparent adsorption capacity, and, therefore, occupy all of the reaction volume (5 mL of AcOH vs. 150 mg of catalyst). To extract reaction products for HPLC analysis, up to 40 mL of DMSO (the best solvent for terephthalic and *p*-toluic acids) was required, that strictly reduced the possibility for the hot filtration and recycling tests. Mn and Co content in the catalyst recycled was measured just after filtration from DMSO solution. Therefore, the metal leaching could take place not only during the reaction, but also at washing by DMSO.

Hence one may conclude, that pores and cavities of MCM-41 and halloysite moieties act as microreactors for the oxidation of *p*-xylene under the reaction conditions. However these pores appear too wide and not able to effectively retain Mn and Co species inside the carrier at filtration and washing as compared with polynuclear μ_3-oxo-bound complexes of Co and Mn, encapsulated in the cavities of Y zeolite [16]. This challenges for the further development and modification of MCM-41/HNT composites to provide more effective metal retention and leaching resistance under the severe reaction conditions, to make possible the repeated use of $Mn^{II}{}_1Co^{II}{}_{10}$@MCM-41/HNT catalyst in *p*-xylene oxidation.

3. Experimental

3.1. Chemicals

The following substances were used as substrates and reference compounds: *p*-xylene 4-$H_3CC_6H_4CH_3$ (pX; Reachim, Purum, Moscow, Russia); *p*-toluic aldehyde 4-$H_3CC_6H_4C(=O)H$ (pTAld) (Aldrich, 97%; Steinheim, Germany); *p*-toluic acid 4-$H_3CC_6H_4C(=O)OH$ (pTAc; Aldrich, 98%; Steinheim, Germany); 4-carboxybenzaldehyde 4-$HO(O=)CC_6H_4C(=O)H$ (4-CBA; Aldrich, 97%; Steinheim, Germany) and terephthalic acid *p*-$HO(O=)CC_6H_4C(=O)OH$ (TPA; Acros Organics, 99+%; Geel, Belgium).

For the synthesis of Mn/Co oxidation catalyst based on MCM-41/HNT composite the following reagents were used: manganese (II) acetate tetrahydrate $(CH_3COO)_2Mn \times 4H_2O$ (Aldrich, ≥99%; Steinheim, Germany), cobaltous (II) acetate tetrahydrate $CH_3COO)_2Co \times 4H_2O$ (Aldrich, ≥98%; Steinheim, Germany) and MCM-41/HNT composite, earlier prepared according to literature procedure [24].

Acetic acid CH_3COOH (Chimmed, Chemical Grade; Moscow, Russia), dimethyl sulfoxide (DMSO) $(H_3C)_2SO$ (Ruschim, Imp; Moscow, Russia) and potassium bromide KBr (Chimmed, Chemical Grade; Moscow, Russia) were used as solvent and additive respectively in the procedure of catalytic experiments.

Double-distilled water, methanol CH_3OH (J.T. Baker, HPLC grade; Gliwice, Poland) and acetonitrile CH_3CN (J.T. Baker, HPLC grade; Gliwice, Poland) were used as solvents and phosphoric acid H_3PO_4 (Component-Reactive, Chemical Grade; Moscow, Russia) was used as a pH regulating additive while conducting HPLC analysis.

3.2. Analyses and Instrumentations

Analysis by transmission electron microscopy (TEM) was carried out using LEO912 AB OMEGA (Carl Zeiss, Jena, Germany) and JEM-2100 JEOL microscopes (Jeol Ltd., Tokyo, Japan) with an electron tube voltage of 100 kV.

The isotherms of nitrogen adsorption/desorption were measured at 77 K on Micromeritics Gemini VII 2390 t instrument (Micromeritics Instrument Corp., Norcross, GA, USA). Before the measurements, the samples were degassed at 350 °C for 4 h. The specific surface area was calculated with the Brunauer–Emmett–Teller (BET) and Langmuir methods applied to the range of relative pressures $P/P_0 = 0.05–0.30$. The pore volume and pore size distributions were determined from the adsorption branches of the isotherms based on the Barrett–Joyner–Halenda (BJH) model.

Weight content of manganese and cobalt in the catalyst was determined by means of atomic emission spectroscopy with inductively coupled plasma (ICP-AES) and X-ray fluorescent spectrometry (XFS). ICP-AES analysis was performed on the IRIS Intrepid II XPL instrument (Thermo Electron Corp., Waltham, MA, USA) in the radial observation modes at wavelengths of 257.6 and 259.4 nm for Mn and 228.6 and 237.9 nm for Co. XFS analysis was conducted on the ARL PERFORM'X spectrometer (Thermo Fisher Scientific, Waltham, MA, USA) with Rh anode and tube diameter of 3 mm in the etalon-free mode with the relative error of 5%.

X-ray photoelectron studies (XPS) were carried out on a Versa Probe II instrument (ULVAC-PHI Inc., Hagisono, Chigasaki, Kanagawa, Japan), equipped with a photo-electronic hemispherical analyzer with retarding potential OPX-150. X-ray radiation of the aluminum anode (Al $K\alpha$ = 1486.6 eV) with a tube voltage of 12 kV, emission current of 20 mA and power of 50 W was used to excite photoelectrons. Photoelectron peaks were calibrated with respect to the carbon C 1s line with a binding energy of 284.8 eV.

Qualitative and quantitative analysis for the products of the *p*-xylene oxidation was carried out by means of high-performance liquid chromatography (HPLC) similar to the procedure, earlier described in literature [56] on the Agilent 1100 Series instrument (Agilent Technologies, Santa Clara, CA, USA) with a Zorbax SB-C_{18} column (5 μm, 2.1 mm × 150 mm) and UV detector. The eluent consisted of methanol (25%), acetonitrile (25%) and water (50%). For better peak resolution, phosphoric acid was added with concentration of 1 mL per 1 L of the eluent. The flow rate was 0.1 mL/min and the injection volume was 0.1 μL. Chromatograms were recorded at 210, 230, 254 and 280 nm simultaneously and analyzed on a computer using Agilent 1100 Software (Agilent Technologies, 2008, Santa Clara, CA, USA). Conversion of *p*-xylene and the yield of each oxidation product were calculated using calibrating curves with exponential approximation.

3.3. The Synthesis of Mn/Co-Containing Oxidation Catalyst Based on the MCM-41/HNT Composite

Deposition of Mn and Co on the MCM-41/HNT composite was performed according to the following procedure [16]. Of MCM-41/HNT composite 1000 mg and 20 mL of distilled water were placed into a 100 mL one-neck round-bottom flask, equipped with a reflux condenser and a magnetic stirrer. Then 20 mg of $Mn(OAc)_2 \times 4H_2O$ (0.08 mmol) and 200 mg of $Co(OAc)_2 \times 4H_2O$ (0.8 mmol) were placed into a chemical beaker and dissolved in 10 mL of distilled water at room temperature while stirring. Then the resulting solution of the pink-purple color was added dropwise to the suspension of MCM-41/HNT composite in water at room temperature while stirring, and additional 10 mL of distilled water was passed through the dropping funnel to wash the residues of $Mn(OAc)_2$ and $Co(OAc)_2$ solution. The reaction was carried out at 60 °C for 12 h. After that the material obtained was centrifuged, washed twice with ethanol for the better removal of water and then dried in the air. The catalyst obtained was isolated as a pale-pink powder, weighing 1065 mg (92%).

ω_{Mn} (XFS): 0.15%.
ω_{Co} (XFS): 1.29%.

XPS (eV): 74.5 (Al_2O_3, Al 2p, 2.5%); 103.5 (SiO_2, Si 2p, 25.9%); 119.5 (Al_2O_3, Al 2s); 154.5 (SiO_2, Si 2s); 284.9 (–CH_2CH_2–, C 1s, 1.27%), 285.9 (–CH_2CH_2N, $H_3CC(=O)$–, C 1s, 3.15%), ($H_3CC(=O)$–, –CH_2N^+, C 1s, 0.82%), 290.1 ($H_3CC(=O)$–, C 1s sat, 0.55%); 400.5 (–CH_2CH_2N, $[NR_4]^+$, N 1s, 0.6%); 530.6 (MnO_x, CoO_x, O 1s, 1.7%), 532.1 ($CH_3C(=O)O$–, $(Al_2O_3)_x$*$(SiO_2)_y$, O 1s, 14.8%), 533.1 (SiO_2, O 1s, 43.0%), 534.6 (O … H_2O, O 1s, 4.6%); 642.5 (MnO, Mn $2p_{3/2}$, 0.08%), 646.9 ($[MnO_4]$ bound, Mn $2p_{3/2}$, 0.02%), 654.0 (MnO, Mn $2p_{1/2}$), ($[MnO_4]$ bound, Mn $2p_{1/2}$); 686.5 (O–Si–F, F 1s, 0.2%); 780.0 (CoO, Co $2p_{3/2}$, 0.05%), 782.5 ($[CoO_6]$ bound, Co $2p_{3/2}$, 0.28%), 785.2 (CoO, $[CoO_6]$ bound, Co $2p_{3/2}$ sat, 0.14%), 788.2 (CoO, $[CoO_6]$ bound, Co $2p_{3/2}$ sat, 0.17%), 791.6 (Co $2p_{3/2}$ sat, 0.06%), 794.6 (CoO, Co $2p_{1/2}$), 798.4 ($[CoO_6]$ bound, Co $2p_{1/2}$), 802.4 (CoO, $[CoO_6]$ bound, Co $2p_{1/2}$ sat), 805.1 (CoO, $[CoO_6]$ bound, Co $2p_{1/2}$ sat), 808.9 (Co $2p_{1/2}$ sat, 0.06%).

3.4. Protocol for the Catalytic Experiments

Oxidation of *p*-xylene in the presence of Mn/Co MCM-41/HNT catalyst was carried out according to the literature procedures [16,29]. Here 150 mg of the catalyst, 4.5 mg of KBr, 0.5 mL of *p*-xylene and 5 mL of acetic acid were placed in a titanium autoclave, equipped with a magnetic stirrer. The autoclave was sealed, filled with oxygen up to a pressure of 2 MPa and placed in an oven with a thermostat control. Then the oxidation reaction was conducted at 150, 200 or 250 °C for 1, 3 or 5 h. After reaction, the autoclave was cooled down to room temperature and depressurized. The reaction products were additionally diluted by DMSO as the best solvent for terephthalic acid and, after the catalyst sedimentation and filtration, analyzed by the HPLC method.

The catalyst activity (TOF = turnover frequency) was calculated as the amount of reacted substrate (ν_{substr}) per mole of metal ($\nu_{(Mn\ +\ Co)}$) per unit of time with account of the yield of each product and number the oxygen atoms added, according to the formula:

$$TOF(O_2) = \frac{\nu_{substr} * (\omega_{pTAld}*1 + \omega_{pTAc}*2 + \omega_{4-CBA}*3 + \omega_{TPA}*4 + \omega_{N/I}*3)}{\nu_{(Mn+Co)} * t},$$

where ω is the yield of the certain product, expressed in the unit fractions, N/I is a not identified product (presumably 4-hydroxymethylbenzoic acid) and *t* is the minimal reaction time, for which the reaction progress is measured.

Each experiment at the same conditions was carried out two or three times, with the results differing by no more than 5% from the corresponding average value. These average values were presented in Tables 2–4. The measurement error did not exceed 5%.

4. Conclusions

Heterogeneous bimetallic Mn/Co-containing catalyst, based on MCM-41/halloysite composite, have been synthesized for the first time and tested in *p*-xylene oxidation under the conditions of the AMOCO process. It was demonstrated, that in the presence of bimetallic heterogeneous catalyst $Mn^{II}{}_1Co^{II}{}_{10}$@MCM-41/HNT and KBr as a free radical source, the quantitative yield of terephthalic acid can be obtained in 3 h at temperature of above 200 °C, and oxygen pressure of 20 atm. The substrate to the Mn + Co ratio was 3.5–4 times higher than that for the traditional homogeneous

$Mn(OAc)_2/Co(OAc)_2$ system, hence proving a high efficacy and superiority of the heterogeneous catalyst, based on the hierarchical material MCM-41/HNT.

The influence of the oxygen pressure, temperature and KBr presence on the catalyst activity and product distribution was investigated. It was established that a decrease in oxygen pressure to 5 atm. resulted in the corresponding decrease of *p*-xylene conversion to 88–89%, with *p*-toluic acid obtained as the major reaction product with the yield up to 63% within 3 h. Decrease in temperature from 200 to 150 °C led to the abrupt downfall in the reaction rate. Herein the conversion did not exceed 40% after 5 h and *p*-toluic aldehyde was the major reaction product. Vice versa, rise in temperature from 200 to 250 °C did not result in the significant increase of the reaction turnover frequency. The presence of KBr was found to be crucial for the effective process of the oxidation of *p*-xylene, whose conversion did not exceed 2%, when KBr was removed from the reaction medium.

It was found that in the presence of $Mn^{II}{}_1Co^{II}{}_{10}$@MCM-41/HNT the further oxidation of *p*-toluic acid to terephthalic acid mostly proceeded through the formation of 4-hydroxymethylbenzoic acid, thus eliminating the stage of undesirable 4-carboxybenzaldehyde. This pathway supposes the diminished role of the Co (II) in the oxidation of *p*-toluic acid and, therefore, allows the elevated substrate to catalyst ratios, but requires high oxygen pressures and Br· radicals as initiators.

Halloysite clay is a cheap and available in thousands tons aluminosilicate, which makes it a prospective nanomaterial for catalysts support. Despite the low resistant to metal leaching under the reaction and separation conditions the heterogeneous catalyst showed a phenomenally high activity in the oxidation of *p*-xylene to terephthalic acid under the conditions of AMOCO industrial process. This $Mn^{II}{}_1Co^{II}{}_{10}$@MCM-41/HNT catalyst based on new hierarchical support with halloysite aluminosilicate nanotubes could be easily scaled up after stability improvement.

Author Contributions: Conceptualization, A.G., V.V. and A.M.; Methodology, A.G. and A.Z.; Software, A.Z. and E.I.; Validation, E.K. and A.M.; Formal analysis, E.I.; Investigation, A.Z. and A.G.; Resources, E.I.; Data curation, A.Z.; Writing—Original draft preparation, A.Z. and A.G.; Writing—Review and editing, A.M. and E.K.; Visualization, A.G. and A.Z.; Supervision, V.V.; Project administration, V.V. and A.M.; Funding acquisition, E.I. All authors have read and agreed to the published version of the manuscript.

Funding: The study was financially supported by the Ministry of Science and Higher Education of the Russian Federation, the unique project identifier is RFMEFI57717X0239.

Acknowledgments: We also thank National University of Science and Technology 'MISIS' (Moscow, Russia) for XPS facilities and Valentine Stytsenko (Gubkin University) and Yusuf Darrat (Louisiana Tech University, USA) for language editing.

Conflicts of Interest: The authors declare no conflict of interest.

References

1. Sheehan, R.J. Terephthalic acid, dimethyl terephthalate, and isophthalic acid. In *Ullmann's Encyclopedia of Industrial Chemistry*; Wiley-VCH Verlag GmbH & Co. KGaA: Weinheim, Germany, 2002.
2. Nazimok, V.F.; Ovchinnikov, V.I.; Potekhin, V.M. *Liquid-Phase Oxidation of Alkyl Aromatic Hydrocarbons*; Chemistry: Moscow, Russia, 1987; p. 240. (In Russian)
3. Li, M.; Niu, F.; Zuo, X.; Metelski, P.D.; Busch, D.H.; Subramaniam, B. A spray reactor concept for catalytic oxidation of *p*-xylenetoproduce high-purity terephthalicacid. *Chem. Eng. Sci.* **2013**, *104*, 93–102. [CrossRef]
4. Partenheimer, W. Methodology and scope of metal/bromide autoxidation of hydrocarbons. *Catal. Today* **1995**, *23*, 69–158. [CrossRef]
5. Saffer, A.; Barker, R.S. Preparation of Aromatic Polycarboxylic Acids. U.S. Patent 2833816 A, 6 May 1958.
6. Saffer, A.; Barker, R.S. Oxidation Chemical Process. U.S. Patent 3089906 A, 14 May 1963.
7. Karakhanov, E.A.; Maksimov, A.L.; Zolotukhina, A.V.; Vinokurov, V.A. Oxidation of *p*-xylene. A review. *Russ. J. Appl. Chem.* **2018**, *91*, 707–727. [CrossRef]
8. Tomás, R.A.F.; Bordado, J.C.M.; Gomes, J.F.P. *p*-xylene oxidation to terephthalic acid: A literature review oriented toward process optimization and development. *Chem. Rev.* **2013**, *113*, 7421–7469. [CrossRef] [PubMed]

9. Ichikawa, Y.; Yamashita, G.; Tokashiki, M.; Yamaji, T. New Oxidation Process for Production of Terephthalic Acid from p-Xylene. *Ind. Eng. Chem.* **1970**, *62*, 38–42. [CrossRef]
10. Fadzil, N.A.M.; Rahim, M.H.A.; Maniam, G.P. A brief review of *para*-xylene oxidation to terephthalic acid as a model of primary C–H bond activation. *Chin. J. Catal.* **2014**, *35*, 1641–1652. [CrossRef]
11. Zuo, X.; Subramaniam, B.; Busch, D.H. Liquid-phase oxidation of toluene and *p*-toluic acid under mild conditions: Synergistic effects of cobalt, zirconium, ketones, and carbon dioxide. *Ind. Eng. Chem. Res.* **2008**, *47*, 546–552. [CrossRef]
12. Sun, W.; Yi, P.; Zhao, L.; Zhou, X. Simplified free-radical reaction kinetics for *p*-xylene oxidation to terephthalic acid. *Chem. Eng. Technol.* **2008**, *31*, 1402–1409. [CrossRef]
13. Zuo, X.; Niu, F.; Snavely, K.; Subramaniam, B.; Busch, D.H. Liquid phase oxidation of *p*-xylene to terephthalic acid at medium-high temperatures: Multiple benefits of CO_2-expanded liquids. *Green Chem.* **2010**, *12*, 260–267. [CrossRef]
14. Hronec, M.; Hrabě, Z. Liquid-phase oxidation of *p*-xylene catalyzed by metal oxides. *Ind. Eng. Chem. Prod. Res. Develop.* **1986**, *25*, 257–261. [CrossRef]
15. Li, Y.; Duan, D.; Wu, M.; Li, J.; Yan, Z.; Wang, W.; Zi, G.; Wang, J. One-step synthesis of 2,5-dihydroxyterephthalic acid by the oxidation of p-xylene over M-MCM-41 (M = Fe, Fe/Cu, Cu) catalysts. *Chem. Eng. J.* **2016**, *306*, 777–783. [CrossRef]
16. Chavan, S.A.; Srinivas, D.; Ratnasamy, P. Selective oxidation of *para*-xylene to terephthalic acid by μ_3-oxo-bridged Co/Mn cluster complexes encapsulated in zeolite–Y. *J. Catal.* **2001**, *204*, 409–419. [CrossRef]
17. Lvov, Y.; Wang, W.; Zhang, L.; Fakhrullin, R. Halloysite clay nanotubes for loading and sustained release of functional compounds. *Adv. Mater.* **2016**, *28*, 1227–1250. [CrossRef] [PubMed]
18. Vinokurov, V.A.; Stavitskaya, A.V.; Chudakov, Y.A.; Ivanov, E.V.; Shrestha, L.K.; Ariga, K.; Darrat, Y.A.; Lvov, Y.M. Formation of metal clusters in halloysite clay nanotubes. *Sci. Technol. Adv. Mater.* **2017**, *18*, 147–151. [CrossRef]
19. Vinokurov, V.A.; Stavitskaya, A.V.; Glotov, A.P.; Novikov, A.A.; Zolotukhina, A.V.; Kotelev, M.S.; Gushchin, P.A.; Ivanov, E.V.; Darrat, Y.; Lvov, Y.M. Nanoparticles formed onto/into halloysite clay tubules: Architectural synthesis and applications. *Chem. Rec.* **2018**, *18*, 858–867. [CrossRef]
20. Vinokurov, V.; Glotov, A.; Chudakov, Y.; Stavitskaya, A.; Ivanov, E.; Gushchin, P.; Zolotukhina, A.; Maximov, A.; Karakhanov, E.; Lvov, Y. Core/shell ruthenium–halloysite nanocatalysts for hydrogenation of phenol. *Ind. Eng. Chem. Res.* **2017**, *56*, 14043–14052. [CrossRef]
21. Glotov, A.; Stavitskaya, A.; Chudakov, Y.; Ivanov, E.; Huang, W.; Vinokurov, V.; Zolotukhina, A.; Maximov, A.; Karakhanov, E.; Lvov, Y. Mesoporous metal catalysts templated on clay nanotubes. *Bull. Chem. Soc. Jpn.* **2019**, *92*, 61–69. [CrossRef]
22. Glotov, A.; Stytsenko, V.; Artemova, M.; Kotelev, M.; Ivanov, E.; Gushchin, P.; Vinokurov, V. Hydroconversion of aromatic hydrocarbons over bimetallic catalysts. *Catalysts* **2019**, *9*, 38. [CrossRef]
23. Glotov, A.P.; Stavitskaya, A.V.; Chudakov, Y.A.; Artemova, M.I.; Smirnova, E.M.; Demikhova, N.R.; Shabalina, T.N.; Gureev, A.A.; Vinokurov, V.A. Nanostructured ruthenium catalysts in hydrogenation of aromatic compounds. *Petrol. Chem.* **2018**, *58*, 1221–1226. [CrossRef]
24. Glotov, A.; Levshakov, N.; Stavitskaya, A.; Artemova, M.; Gushchin, P.; Ivanov, E.; Vinokurov, V.; Lvov, Y. Templated self-assembly of ordered mesoporous silica on clay nanotubes. *Chem. Commun.* **2019**, *55*, 5507–5510. [CrossRef]
25. Lvov, Y.; Panchal, A.; Fu, Y.; Fakhrullin, R.; Kryuchkova, M.; Batasheva, S.; Stavitskaya, A.; Glotov, A.; Vinokurov, V. Interfacial self-assembly in halloysite nanotube composites. *Langmuir* **2019**, *35*, 8646–8865. [CrossRef] [PubMed]
26. Glotov, A.; Levshakov, N.; Vutolkina, A.; Lysenko, S.; Karakhanov, E.; Vinokurov, V. Aluminosilicates supported La containing sulfur reduction additives for FCC catalyst: Correlation between activity, support structure and acidity. *Catal. Today* **2019**, *329*, 135–141. [CrossRef]
27. Glotov, A.P.; Levshakov, N.S.; Vutolkina, A.V.; Lysenko, S.V.; Gushchin, P.A.; Vinokurov, V.A. Bimetallic sulfur reduction additives based on alumosilicate of Al-MCM-41 type for cracking catalysts: Desulfurazing activity vs. ratio of components in a support. *Russ. J. Appl. Chem.* **2019**, *92*, 562–568. [CrossRef]
28. Karakhanov, E.; Akopyan, A.; Golubev, O.; Anisimov, A.; Glotov, A.; Vutolkina, A.; Maximov, A. Alkali earth catalysts based on mesoporous MCM-41 and Al-SBA-15 for sulfone removal from middle distillates. *ACS Omega* **2019**, *4*, 12736–12744. [CrossRef] [PubMed]

29. Ghiaci, M.; Mostajeran, M.; Gil, A. Synthesis and characterization of Co–Mn nanoparticles immobilized on a modified bentonite and its application for oxidation of *p*-xylene to terephthalic acid. *Ind. Eng. Chem. Res.* **2012**, *51*, 15821–15831. [CrossRef]
30. Carver, J.C.; Schweitzer, G.K.; Carlson, T.A. Use of X-Ray Photoelectron Spectroscopy to Study Bonding in Cr, Mn, Fe, and Co Compounds. *J. Chem. Phys.* **1972**, *57*, 973–981. [CrossRef]
31. Franzen, H.F.; Umana, M.X.; McCreary, J.R.; Thorn, R.J. XPS spectra of some transition metal and alkaline earth monochalcogenides. *J. Solid State Chem.* **1976**, *18*, 363–368. [CrossRef]
32. Tan, B.J.; Klabunde, K.J.; Sherwood, P.M.A. XPS studies of solvated metal atom dispersed (SMAD) catalysts. Evidence for layered cobalt-manganese particles on alumina and silica. *J. Am. Chem. Soc.* **1991**, *113*, 855–861. [CrossRef]
33. McIntyre, N.S.; Cook, M.G. X-ray photoelectron studies on some oxides and hydroxides of cobalt, nickel, and copper. *Anal. Chem.* **1975**, *47*, 2208–2213. [CrossRef]
34. Mc Intyre, N.S.; Johnston, D.D.; Coatsworth, L.L.; Davidson, R.D.; Brown, J.R. X-ray photoelectron spectroscopic studies of thin film oxides of cobalt and molybdenum. *Surf. Interface Anal.* **1990**, *15*, 265–272. [CrossRef]
35. Kim, K.S. X-ray-photoelectron spectroscopic studies of the electronic structure of CoO. *Phys. Rev. B* **1975**, *11*, 2177–2185. [CrossRef]
36. Umezawa, Y.; Reilley, C.N. Effect of argon ion bombardment on metal complexes and oxides studied by x-ray photoelectron spectroscopy. *Anal. Chem.* **1978**, *50*, 1290–1295. [CrossRef]
37. Oku, M. X-ray photoelectron spectra of $KMnO_4$ and K_2MnO_4 fractured in situ. *J. Electron Spectrosc. Relat. Phenom.* **1995**, *74*, 135–148. [CrossRef]
38. Brown, D.G.; Weser, U. XPS spectra of spin-triplet cobalt (III) complexes. *Z. Nat. B* **1979**, *34*, 1468–1470. [CrossRef]
39. Nefedov, V.I.; Baranovskii, I.B.; Molodkin, A.K.; Omuralieva, V.O. X-ray photoelectron study of cobalt compounds. *Russ. J. Inorg. Chem.* **1973**, *18*, 1295. (In Russian)
40. Nefedov, V.I.; Gati, D.; Dzhurinskii, B.F.; Sergushin, N.P.; Salyn, Y.V. X-ray photoelectron study of several elements oxides. *Russ. J. Inorg. Chem.* **1975**, *20*, 2307. (In Russian)
41. Nefedov, V.I.; Firsov, M.N.; Shaplygin, I.S. Electronic structures of $MRhO_2$, MRh_2O_4, $RhMO_4$ and Rh_2MO_6 on the basis of X-ray spectroscopy and ESCA data. *J. Electron Spectrosc. Relat. Phenom.* **1982**, *26*, 65–78. [CrossRef]
42. Molinaro, F.S.; Little, R.G.; Ibers, J.A. Oxygen binding to a model for the active site in cobalt-substituted hemoglobin. *J. Am. Chem. Soc.* **1977**, *99*, 5628–5632. [CrossRef]
43. Emara, A.A.A.; Ali, A.M.; El-Asmy, A.F.; Ragab, E.-S.M. Investigation of the oxygen affinity of manganese (II), cobalt (II) and nickel (II) complexes with some tetradentate Schiff bases. *J. Saudi Chem. Soc.* **2011**, *18*, 762–773. [CrossRef]
44. Beamson, G.; Briggs, D. *High Resolution XPS of Organic Polymers: The Scienta ESCA300 Database*; John Wiley & Sons: Chichester, UK, 1992; 295p.
45. Briggs, D.; Beamson, G. Primary and secondary oxygen-induced C 1s binding energy shifts in X-ray photoelectron spectroscopy of polymers. *Anal. Chem.* **1992**, *64*, 1729–1736. [CrossRef]
46. Yuan, H.; Fang, X.; Ma, Q.; Mao, J.; Chen, K.; Chen, Z.; Li, H. New mechanistic insight into the aerobic oxidation of methylaromatic compounds catalyzed by Co–Mn–Br and its applications. *J. Catal.* **2016**, *339*, 284–291. [CrossRef]
47. Wang, Q.; Li, X.; Wang, L.; Cheng, Y.; Xie, G. Effect of water content on the kinetics of *p*-xylene liquid-phase catalytic oxidation to terephthalic acid. *Ind. Eng. Chem. Res.* **2005**, *44*, 4518–4522. [CrossRef]
48. Li, M.; Niu, F.; Busch, D.H.; Subramaniam, B. Kinetic investigations of *p*-xylene oxidation to terephthalic acid with a Co/Mn/Br catalyst in a homogeneous liquid phase. *Ind. Eng. Chem. Res.* **2014**, *53*, 9017–9026. [CrossRef]
49. Cheng, Y.; Li, X.; Wang, L.; Wang, Q. Optimum ratio of Co/Mn in the liquid-phase catalytic oxidation of *p*-xylene to terephthalic acid. *Ind. Eng. Chem. Res.* **2006**, *45*, 4156–4162. [CrossRef]
50. Li, K.; Li, S. $CoBr_2$-$MnBr_2$ containing catalysts for catalytic oxidation of *p*-xylene to terephthalic acid. *Appl. Catal. A Gen.* **2008**, *340*, 271–277. [CrossRef]

51. Liu, H.; Chen, G.; Jiang, H.; Li, Y.; Luque, R. From alkyl aromatics to aromatic esters: Efficient and selective C–H activation promoted by a bimetallic heterogeneous catalyst. *ChemSusChem* **2012**, *5*, 1892–1896. [CrossRef] [PubMed]
52. Kesavan, L.; Tiruvalam, R.; Ab Rahim, M.H.; bin Saiman, M.I.; Enache, D.I.; Jenkins, R.L.; Dimitratos, N.; Lopez-Sanchez, J.A.; Taylor, S.H.; Knight, D.W.; et al. Solvent-free oxidation of primary carbon-hydrogen bonds in toluene using Au-Pd alloy nanoparticles. *Science* **2011**, *331*, 195–199. [CrossRef] [PubMed]
53. Tbibitt, J.M.; Gong, W.H.; Schammel, W.P.; Hepfer, R.P.; Adamian, V.; Brugge, S.P.; Metelski, P.D.; Zhou, C. Process for the Production of Aromatic Carboxylic Acids in Water. WO Patent 2007133976 A2, 22 November 2007.
54. Tashiro, Y.; Iwahama, T.; Sakaguchi, S.; Ishii, Y. A new strategy for the preparation of terephthalic acid by the aerobic oxidation of *p*-xylene using *N*-hydroxyphthalimide as a catalyst. *Adv. Synth. Catal.* **2001**, *343*, 220–225. [CrossRef]
55. Saha, B.; Koshino, N.; Espenson, J.H. *N*-hydroxyphthalimides and metal cocatalysts for the autoxidation of *p*-xylene to terephthalic acid. *J. Phys. Chem. A* **2004**, *108*, 425–431. [CrossRef]
56. Xiao, Y.; Zhang, X.Y.; Wang, Q.B.; Tan, Z.; Guo, C.C.; Deng, W.; Liu, Z.G.; Zhang, H.F. The preparation of terephthalic acid by solvent-free oxidation of *p*-xylene with air over T(*p*-Cl)PPMnCl and $Co(OAc)_2$. *Chin. Chem. Lett.* **2011**, *22*, 135–138. [CrossRef]

© 2019 by the authors. Licensee MDPI, Basel, Switzerland. This article is an open access article distributed under the terms and conditions of the Creative Commons Attribution (CC BY) license (http://creativecommons.org/licenses/by/4.0/).

Article

The Influence of Preparation Method on the Physicochemical Characteristics and Catalytic Activity of Co/TiO_2 Catalysts

John Vakros

Department of Chemistry, University of Patras, Rion, GR26504 Patras, Greece; vakros@chemistry.upatras.gr

Received: 23 December 2019; Accepted: 6 January 2020; Published: 7 January 2020

Abstract: Two Co/TiO_2 catalysts with 7% CoO/g loading were prepared using equilibrium deposition filtration and the dry impregnation method. The two catalysts were characterized with various physicochemical techniques and tested for the degradation of sulfamethaxazole (SMX) using sodium persulfate (SPS) as the oxidant. It was found that the two catalysts exhibit different physicochemical characteristics. The equilibrium deposition filtration (EDF) catalyst had a higher dispersion of cobalt phase, more easily reduced Co(III) species, and a higher ratio of Co(III)/Co(II) species. The interactions between Co-deposited species and the titania surface were monitored with diffuse reflectance spectroscopy in all the preparation steps, and it was found that they increased during drying and calcination, while EDF favored the formation of surface species with strong interactions with the support. Finally, the EDF catalyst was more active for the degradation of sulfamethaxazole due to its better physicochemical characteristics.

Keywords: cobalt catalyst; titania; diffuse reflectance spectroscopy; sulfamethaxazole; persulfates; point of zero charge

---

1. Introduction

Cobalt-supported catalysts are quite promising for many reactions of environmental and industrial interest such as volatile organic compounds oxidation [1], CO_2 hydrogenation [2], the Fischer–Tropsch process [3,4], and electrocatalysis [5].

In the last years, many studies dealt with the cobalt catalysts supported on titania [6–9]. In most of them, extensive physicochemical characterization was performed in order to combine the catalytic activity with the physicochemical characteristics. At this point, it should be stated that, generally and for Co catalysts specifically, the physicochemical characteristics and consequently the catalytic activity are strongly dependent on the preparation method and the precursor-supported species [10–14].

Thus, there are many studies dealing with the preparation of Co/TiO_2 catalysts with different methods, although the most common method used for the preparation of supported Co catalysts is the dry impregnation (DI) method. This is because of its simplicity and low cost, as the other methods used are quite complicated and costly. Using DI, the speciation of Co-supported species is usually determined by the loading of the Co phase and the TiO_2 properties such as the specific surface area, surface groups, and point-of-zero charge. The deposition mechanism is mainly the bulk precipitation during the drying step, while only a small portion of Co species adsorbs onto the titania surface. It is difficult to adjust the impregnation parameters during dry impregnation and consequently to control the mechanism of deposition of active species and the interactions of the supported phase.

One very attractive method for the preparation of catalysts is the equilibrium deposition filtration method (EDF) [15–17]. This method, compared to the other non-conventional preparation methods used, is simple and it does not need expensive reagents and infrastructure. Using this method, a quite large volume of dilute solution of the precursor compound is used. An amount of support is added to the solution. The suspension is allowed to reach equilibrium under well-defined and controlled impregnation parameters. During equilibrium, a quantity of the precursor ion is deposited onto the support surface. The ion deposition is finished when equilibrium is reached; then, filtration is applied for the collection of the solid and the removal of the non-deposited precursor species. The deposition of the active species can be controlled using proper conditions such as the pH, temperature, initial concentration, etc. There is a variety of different deposition modes such as adsorption, surface reaction, interfacial deposition, etc., depending on the impregnation parameters [11].

Compared to DI, EDF gives us the ability to adjust the impregnation parameters and the interactions of the supported phase with the support surface, and therefore to control the final speciation of the deposited species and the catalytic activity, as it was shown in many studies [1,11,15,16,18,19].

An extensive study about the preparation of Co/TiO_2 catalysts using EDF was performed by Lycourghiotis et al. [20], where the interfacial chemistry of the impregnation step was fully monitored. In that study, it was found that the deposition mechanism involves the formation of mononuclear/oligonuclear inner sphere complexes upon deposition of the $[Co(H_2O)_6]^{2+}$ ions at the "titania/electrolytic solution" interface. Depending on the Co(II) surface concentration, mononuclear complexes are formed, almost exclusively, at low and medium Co(II) surface concentrations. Binuclear and three-nuclear inner sphere complexes are formed, in addition to the aforementioned mononuclear ones, at relatively high Co(II) surface concentrations, while the deposition followed by H^+ released from the solid to bulk solution lowered the pH solution. Finally, it was found that impregnation pH was the most important parameter for the deposition of Co(II) species. Higher pH resulted in a higher amount of Co(II) deposited onto the titania surface; however, pH higher than 8 should be avoided, as Co(II) precipitates either as $Co(OH)_2$ or $Co(OH)NO_3$ in bulk solution. That study focused on the impregnation step and deposition mechanism of Co(II) onto the titania surface. Co/TiO_2 samples using EDF have not been prepared and applied to catalytic tests, which is due to the low Co content.

The limited loading of Co onto the TiO_2 surface is the only disadvantage of the EDF method. The Co loading is low and less than 3% CoO/g of the catalyst, even under the most favorable conditions. This is probably due to the limited solubility of Co^{2+} ions in pH >8. In addition, the deposition mechanism for the Co(II) species onto the titania surface was exclusively through the formation of inner sphere complexes between monomeric or oligomeric Co species and surface–OH groups, and it did not involve surface precipitation. In the case of alumina, surface precipitation can be achieved in the interfacial region in lower pH values than the values required for bulk precipitation, which usually leads to higher deposited amounts of Co and more active catalysts [17,21].

The aim of the present work is the preparation of Co/TiO_2 catalysts with higher loading with two different methods, EDF and DI, and the determination of how the preparation method alters the physicochemical characteristics and the catalytic activity for the degradation of sulfamethaxazole (SMX) with sodium persulfate (SPS) as the oxidant of the prepared catalysts.

In the present work, the EDF method was modified in order to increase the Co loading. Specifically, in the impregnation solution, the pH increased with the addition of a base to a value of 8.5, where extensive hydrolysis and polymerization between Co(II) ions and finally, the precipitation of Co(II) species occurred. The pH remained about 8.5 while Co(II) species precipitated. The amount of base added is about half of the amount required stoichiometrically for the complete precipitation of the Co(II) ions. Under these conditions, only a part of Co(II) precipitates and the rest of the Co(II) ions are dissolved in the suspension, with a higher degree of hydrolysis and polymerization. The pH is still 8.5, which is higher than the one used for the deposition of Co(II) ions in the previous study. This higher pH will be beneficial for the deposited loading of Co(II) species, probably resulting in higher deposited amounts of Co onto titania. In addition, under these conditions, the deposited Co(II) surface complexes onto the surface of titania are expected to have more than 3 Co atoms in contrast with the binuclear or three-nuclear inner sphere complexes formed at lower Co surface coverage.

2. Results

2.1. Preparation of the Catalyst

The solution of Co^{2+} with a concentration of 0.03 M was placed in a double-walled beaker at a constant temperature of 25 °C. The pH was increased to about 8.5 using 2 M of NH_3 solution. The amount of NH_3 was half of the stoichiometric required for the complete precipitation of Co ions. As a result of the limited solubility of Co ions in high pH, the hydroxide ($Co(OH)NO_3$ or $Co(OH)_2$) was precipitated according to equations:

$$Co^{2+}(aq) + NO_3^{-}(aq) + OH^{-}(aq) \leftrightarrows Co(OH)NO_3(s) \tag{1}$$

$$Co^{2+}(aq) + 2\,OH^{-}(aq) \leftrightarrows Co(OH)_2(s). \tag{2}$$

The precipitation was followed by color changes from pink to blue–green. At that point, 1.0 g of TiO_2 was added into the suspension. The surface of TiO_2 was negatively charged in that pH and started to adsorb $Co^{2+}(aq)$ ions. The Co^{2+} ions concentration decreased due to the adsorption process, and Equations (1) and (2) shifted to the left. The positive charge in the interfacial region and the surface of TiO_2 increased due to the accumulation of Co^{2+} ions. The surface reduced the increasingly positive charge by releasing H^+ to the bulk solution. This process resulted in a decrease in pH, and this also shifted Equations (1) and (2) to the left. As a result, the precipitated Co phase acted as a reservoir of Co^{2+} ions to be deposited onto the TiO_2 surface. This Co(II) deposition is usually pH dependable, and generally, it is higher in high pH values. After 5 days of impregnation, the pH is lower than 7.5, and the precipitate Co phase was dissolved. Then, the suspension was filtrated, and the Co/TiO_2 was collected. The supernatant was pink, proving the dissolution of the precipitate Co phase, while the solid Co/TiO_2 had a brown color. The adsorbed Co was measured spectophotometrically, and it was found to be 7% CoO/g of catalyst. On the other hand, the maximum Co content obtained using the standard EDF method was not exceeding 3% under optimal conditions. That was the result of the limited pH used (lower than 7). The preparation route can be seen in Figure 1.

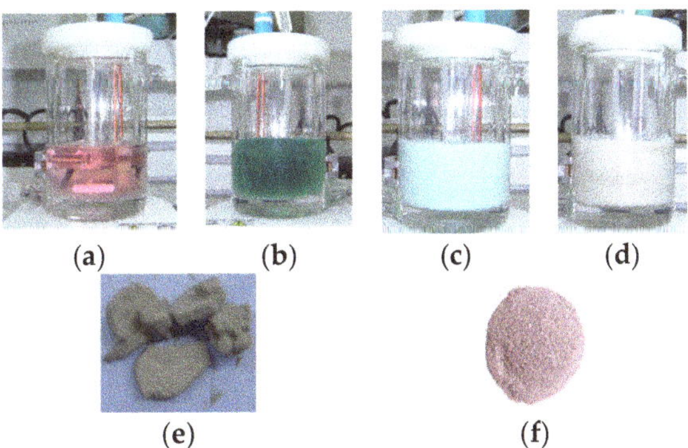

Figure 1. (**a**) Impregnation solution of $Co(H_2O)_6^{2+}$, (**b**) after base addition and pH = 8.5, (**c**) after the addition of TiO_2, (**d**) after deposition for 5 days, (**e**) image of equilibrium deposition filtration method (EDF) catalyst after filtration, and (**f**) image of dry impregnation (DI) catalyst after impregnation.

One other catalyst with the same Co loading was prepared using the DI method. Usually, DI is the most common method to prepare Co-supported catalysts, and the prepared catalysts exhibit high stability and activity.

After the preparation of the catalyst with dry impregnation, the two samples had significant differences, starting from the color of each sample. The DI sample was pink due to the octahedral

$Co(H_2O)_6^{2+}$ complex, indicating that the Co phase is mainly the precipitate $Co(H_2O)_6(NO_3)_2$ phase with low interactions between Co and the surface, while the EDF sample had a brown color suggesting that the local environment of Co(II) was different than the $Co(H_2O)_6^{2+}$. The color change and the differences in the local environment of Co(II) species could be attributed to the partial substitution of water molecules ligands with either O–Ti or O–Co groups. Thus, strong interactions were present between Co ions and surface groups or among Co ions.

2.2. *Physicochemical Characterization with Diffuse Reflectance Spectroscopy (DRS) before Calcination Step*

For the elucidation of the interactions between the Co phase and TiO_2 surface, the diffuse reflectance spectra of the two catalysts were collected after impregnation (wet samples, Figure 2a) and after drying (dry samples, Figure 2b). An inspection of Figure 2a clearly shows the significant differences between the two samples. This was expected, since the two samples had different colors after the impregnation step. Generally, there are three broad peaks in the spectrum. The first peak was centered at about 350 nm, the second was centered at about 500 nm, and a wide broad peak was centered at about 650 nm. In addition, the F(R), which was an analogue to the absorbance of the sample, was higher in the case of the EDF sample, especially in the first and third peaks. Taking into account that the two samples had the same Co loading, the differences in F(R) can be related to the dispersion of the Co phase. It provides evidence that the deposition of the Co phase resulted in higher dispersion in the case of the EDF sample. The first peak, which is located at about 350 nm, is due to the charge transfer between Ti–O–Co, and it can serve as a measurement of the interactions between the surface groups and the Co phase. This peak was absent in the DR spectrum of the $Co(H_2O)_6^{2+}$ ($Co(H_2O)_6(NO_3)_2$), pointing out that this was generated from the interactions between the surface –OH groups and Co ions.

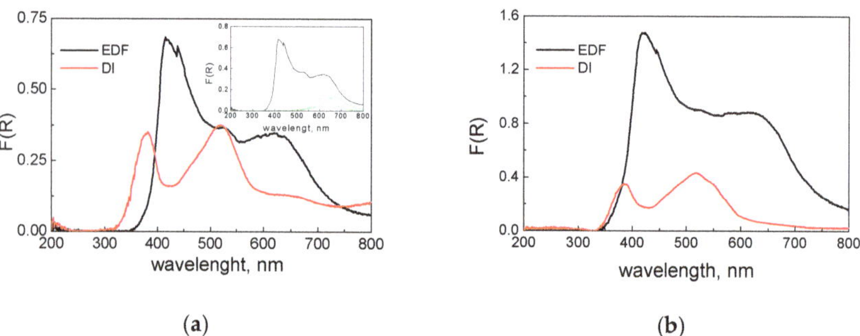

(**a**) (**b**)

Figure 2. Diffuse reflectance spectra of wet (**a**) and dried (**b**) samples.

Generally, this peak has been observed in many cases where a transition metal ion (either as oxyanion, MO_x^{z-}, or hydrated cation, $M(H_2O)_x^{z+}$) deposited on the TiO_2 surface and was attributed to the charge transfer between Ti–O–M bonds [18–20,22,23]. The intensity of the peaks denotes that the interactions were higher in the case of the EDF sample, while the shift observed in lower energy supports this finding. The peak centered at about 500 nm was due to the Co–(OH_2) bonds, and it can be seen in the $Co(H_2O)_6(NO_3)_2$ spectrum. It is characteristic of the octahedral symmetry of Co aqua complexes. The third peak was very interesting, because it features the partial substitution of the H_2O molecules in the $[Co–(OH_2)_6]^{2+}$ complex. The peak can be deconvoluted in two other peaks (inset of Figure 2a), which are centered at about 625 and 680 nm. The first peak is assigned to Co–O–Ti, while the second peak is due to Co–O–Co polymerization in accordance with previous works for Co deposition on an Al_2O_3 surface [21,24].

The same three broad peaks can be seen after drying in Figure 2b. The main difference was the intensity of these peaks. Due to the heat provided during drying, the peaks were more intense, indicating that the interactions were increased as a result of the release of water molecules. A more pronounced increment is evident in the EDF sample. This can be attributed to the better dispersion of the Co phase in the EDF sample. In the EDF case, the main mechanism of the Co phase deposition is the adsorption/surface reaction of Co ions with the –OH surface groups. During drying and water evaporation, the Co phase and the surface groups can interact further, and the corresponding peak is more intense. On the other hand, in the DI case, the main mechanism is the bulk precipitation of the Co phase onto the surface of TiO_2 resulting in agglomerates with higher dimensions. This means that a significant amount of Co phase precipitates onto the previously deposited Co phase, and it is not in contact with the surface –OH groups. The majority of this part was not connected with the surface groups, and during drying, it could not interact with the TiO_2 surface. As a result, the increment of interactions and consequently the intensity of the peak at 400 nm was not significant.

2.3. *Physicochemical Characterization of the Calcined Samples*

2.3.1. Measurement of Specific Surface Area (SSA)

The deposition of 7% CoO on TiO_2 did not significantly alter the SSA of the two samples. The SSA of TiO_2 was found to be equal to 53 m^2/g, while the EDF sample had an SSA equal to 50 m^2/g, and the DI sample had an SSA equal to 45 m^2/g.

The pore diameter distribution (Figure 3) showed small changes for the two samples. DI created a new group of mesoporous with a mean diameter of 28 nm. The EDF impregnation altered the porosity of TiO_2. Specifically, the TiO_2 pores shifted to a lower mean diameter, from 87 nm to 68 nm. This was probably a partial pore blocking due to the deposition of the Co phase onto the surface of TiO_2, which also exhibited a significant increase of porosity, which was probably due to the generation of new pores after Co phase deposition.

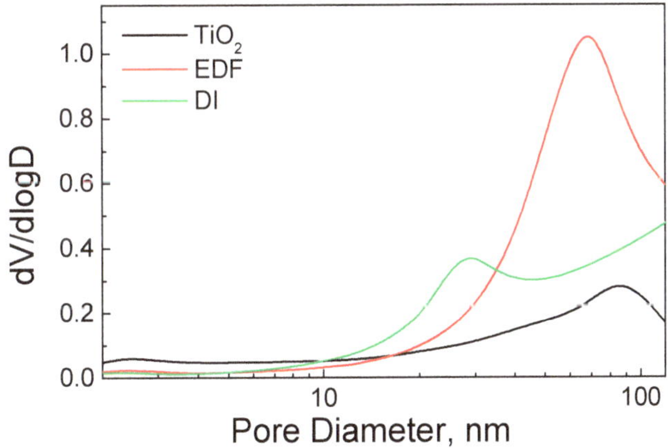

Figure 3. Pore volume distribution of the calcined samples and pure TiO_2.

2.3.2. Diffuse Reflectance Spectra

The samples were calcined at 400 °C. The supported Co phase was transformed to Co_3O_4, as it can be seen in Figure 4. According to the literature, the Co in the Co_3O_4 oxide can be found either in octahedral symmetry (peaks at 425 nm and 710 nm) or in tetrahedral symmetry (the triplet centered at about 600 nm) [1]. As it can be seen in Figure 4, the spectra of the two samples were similar to that of Co_3O_4 oxide. The higher intensity found for the DI sample suggested an increased formation of Co_3O_4 oxides for the DI sample compared to the EDF sample. In addition, the ratio of the peaks at 710 nm and 450 nm was different between the two samples. The DI sample had a very close value to unsupported Co_3O_4, while the value for the EDF sample was higher. These were the results of the different deposition mechanisms and the different speciation of the Co phase before the calcination step. Taking into account that the peak at 710 nm is due to Co(III) ions, it can be claimed that the EDF samples exhibited higher amounts of the Co(III) oxide phase.

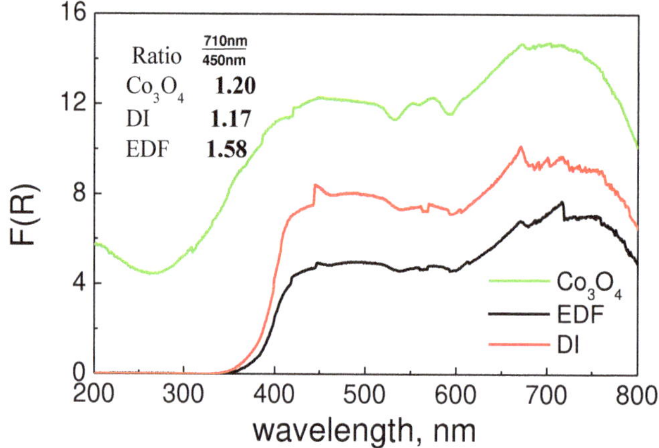

Figure 4. Diffuse reflectance (DR) spectra of the calcined samples.

2.3.3. X-ray Diffraction, XRD

The XRD patterns of the two samples (Figure 5) exhibit only the characteristic XRD peaks of TiO_2 P25. The peaks of the Co_3O_4 supported phase can hardly be observed. Only one corresponding to (311) can be seen at 37°. This indicates that the Co_3O_4-supported particles are either highly dispersed on TiO_2 or the particles are too small to be detected with XRD. Similar results can be found in the literature for catalysts with 10% loading [25–28].

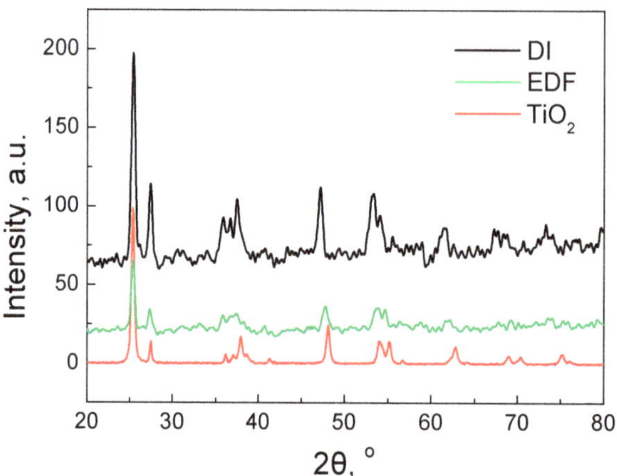

Figure 5. XRD patterns of EDF and DI catalysts and TiO_2 support.

2.3.4. X-ray Photoelectron Spectroscopy, XPS

The X-ray photoelectron spectra for Ti 2p, O 1s, and Co 2p are presented in Figure 6. The two spectra had significant differences as a result of the different deposition mechanism. The region of Ti2p for the DI catalyst showed two symmetrical peaks, which were centered at 459.4 and 465.0 eV.

The shape, the position, and the peak separation (5.6 eV) were very close to the Ti(IV) species in TiO_2 nanoparticles. The spectrum of the EDF catalysts was more complicated. The two peaks were broader. In addition, each peak had a shoulder in lower binding energy (BE). The deconvolution of those peaks clearly showed the contribution of other Ti species in the peaks. These Ti species were located in lower values of BE: 457.0 eV and 462.3 eV respectively. These lower values in BE were probably due to the interactions with Co–O species adsorbed onto the TiO_2 surface groups and the electron charge transfer to Ti atoms in accordance with the DRS results.

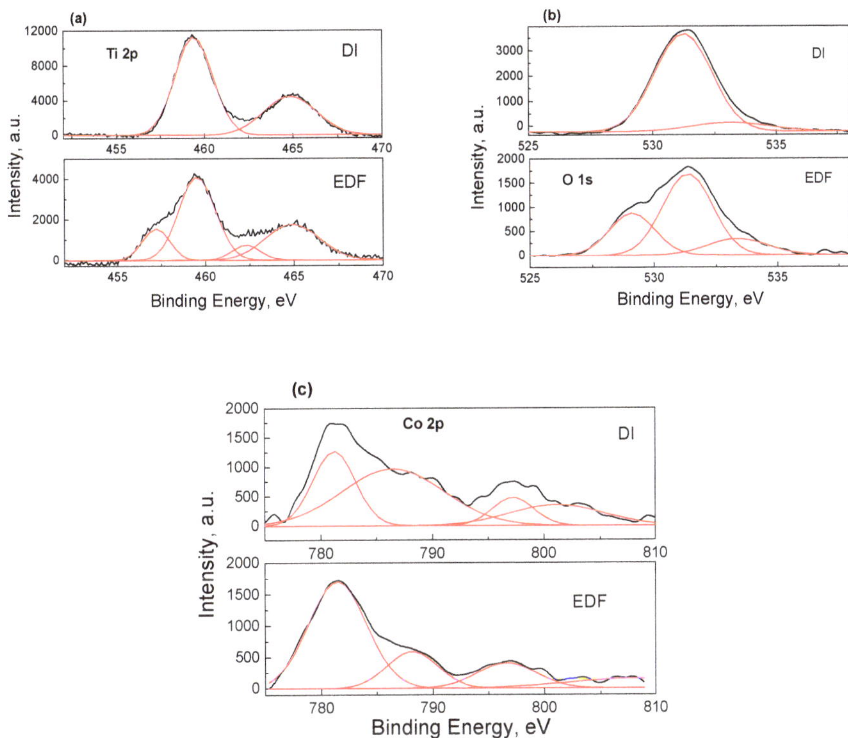

Figure 6. XPS spectra of Ti 2p (**a**), O 1s (**b**), and Co 2p (**c**) photoelectrons for the samples prepared.

The O 1s photoelectron spectrum was also different for the two samples. The DI sample showed one well-formed symmetrical peak centered at 531.4 eV and a shoulder at 533.4 eV. The first peak was characteristic of O in TiO_2, while the second peak was assigned to the surface –OH groups. The O 1s spectrum for the EDF catalyst exhibited one additional new peak centered at 529.1 eV. The lower BE value indicated the increasing of electron cloud density of the lattice oxygen because of the deposition of Co species.

Figure 6c shows the Co 2p spectra of the two samples. There were two main peaks centered at 781.3 and 797.5 eV for DI and 781.5 and 796.4 for the EDF sample and two satellite peaks centered at 786.4 and 801.4 for DI and 788.4 and 807.7 for EDF. The first peak assigned to Co $2p_{3/2}$ and the second peak was assigned to Co $2p_{1/2}$. It should be pointed out that it is rather difficult to determine the oxidation state of Co only by the peak position. Other parameters such as the shape of the satellites and the energy gap between the satellites and the main lines are used to discriminate between the different oxidation states of Co. Generally, the satellite peaks for Co(II) are located at BEs lower than 10 eV with respect to the main peak, while the satellite peaks for Co(III) are detected at BEs higher than 10 eV. For the two samples studied, the satellites peaks were located in higher BEs for the EDF sample, suggesting that higher amounts of Co(III) ions were present on the surface of the EDF sample. Additionally, peaks exhibiting a shoulder are characteristic of Co(II) species in a high spin state, while the diamagnetic low-spin Co(III) ion does not show shake-up structures [25]. In this study, the shoulders were relatively lower in the case of the EDF catalyst, meaning that the Co(III) surface concentration was higher.

On the other hand, higher values of BE for the EDF sample denoted that the Co species interact better with the TiO_2 surface. This suggests that the Co dispersion was higher in the EDF catalyst. Indeed, the atomic ratio Co/Ti for the two samples was found to be 0.11 and 0.20 for DI and EDF respectively, which was almost twice that of the EDF catalyst, proving the better dispersion of the Co phase for the EDF catalyst.

2.3.5. Temperature Programmed Reduction (TPR) with H_2

The TPR profiles of the DI and EDF catalysts are presented in Figure 7. The unsupported Co_3O_4 (inset of Figure 7) usually exhibits one intense peak centered at 454 °C and a shoulder at 369 °C, which is characteristic of the Co(III) species. These peaks were also present in the TPR profile of the DI sample, although the main peak of the DI sample was located at much higher temperature, at about 607 °C. This peak was due to Co(II) species interacting with the TiO_2 surface. The TPR profile for the EDF sample presented three reduction peaks. The first one was located at 373 °C and denotes the Co(III) species of the EDF catalysts. The peak was well formed and of high intensity, suggesting that this catalyst had a significant higher amount of Co(III) species in accordance with the DRS and XPS results. The other two peaks were located at 450 °C and 553 °C, which were both very close to the temperatures of the peaks of the DI samples.

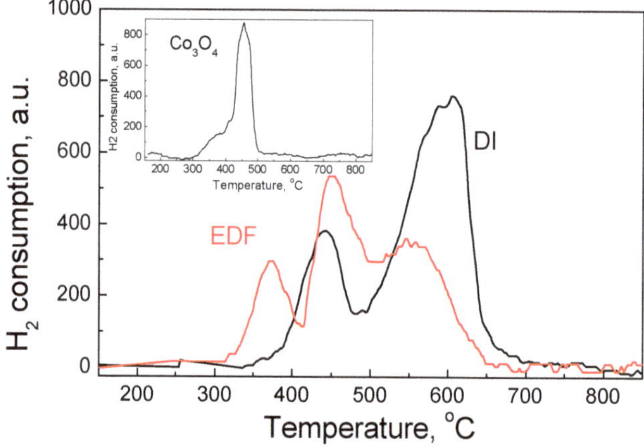

Figure 7. Temperature-programmed reduction (TPR) profiles for the samples studied, inset: the TPR profile for Co_3O_4.

2.3.6. Acid Base Behavior of the Prepared Catalysts

The titration curves of TiO_2 and two catalysts, EDF and DI, are presented in Figure 8. In this figure, there is also the titration curve of the blank solution. The section point of the curve of the blank solution with each of the suspension curves denotes the point of zero charge (pzc) of each solid according to the potentiometric mass titrations (PMT) method.

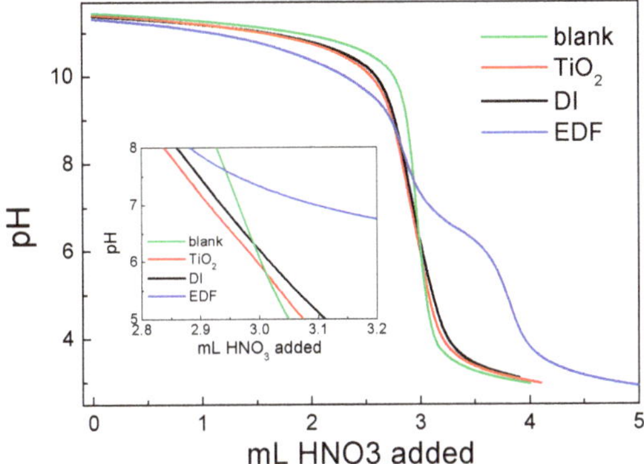

Figure 8. Potentiometric mass titrations for the prepared catalysts, TiO_2 and the blank solution. Inset: the section points of titration curves of each suspension with blank solution, corresponding to the point of zero charge (pzc) of each solid.

As it can be seen from the inset of Figure 8, the pzc for TiO_2 was equal to 5.9. The pzc values for the two catalysts were higher: 6.4 for the DI sample and 7.6 for the EDF sample. As it can be seen from the EDF curve, there was a significant consumption of H^+ at pH values lower than 6. This is evidence that the deposition of Co(II) species occurs through inner sphere complexes and there are significant interactions between the TiO_2 surface and Co-supported phase. The Co(II) deposition process was not pH-independent, but the surface groups were involved. Thus, the acid–base behavior of the catalyst was changed, at least in an aqueous suspension.

The above can be seen more clearly in Figure 9, where the H^+ consumptions as a function of pH suspension were presented for the two catalysts and TiO_2 and Figure 10, where the differential curves of the H^+ consumptions were presented, respectively.

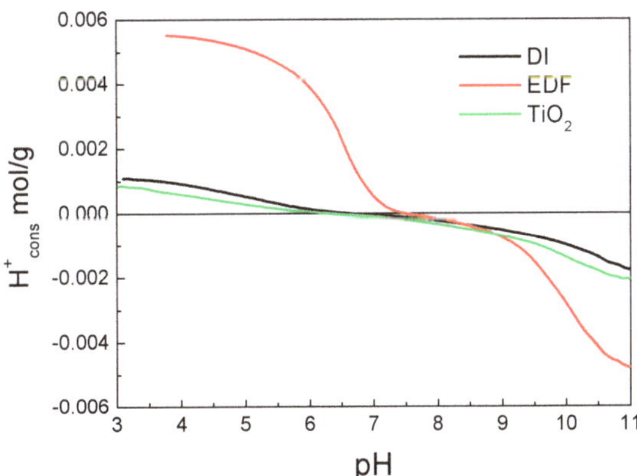

Figure 9. H^+ consumption curves for the three samples as a function of pH suspension.

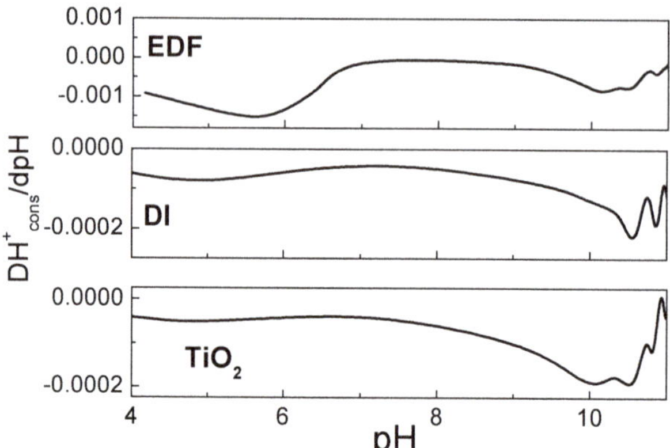

Figure 10. Differential curves of the H^+ consumption for each sample studied.

Indeed, for the EDF sample, a significant amount of H^+ was consumed at a pH lower than 6. This was not expected for the TiO_2, as it is known from the literature that the surface –OH groups do not appear in this pH [27]. Under realistic conditions, where the effect of a double layer is significant, there are two peaks for the TiO_2 centered at about 10 and lower than 3.6, respectively [28]. The peak at a pH of about 10 can be seen in the differential curve for TiO_2 in Figure 10. For the EDF sample, there is a new peak at about 5.5, which can be attributed either to the cobalt phase or to the shift in higher pH of the lower peak of the TiO_2 surface groups. This peak was also present in the DI sample, although it was less intense.

2.4. Catalytic Activity

The prepared catalysts were tested for the degradation of SMX using SPS as the oxidant. This process is part of the advanced oxidation processes (AOPs), which are based on the oxidative power of specific chemicals. In this case, the catalyst activates the SPS to form SO_4^- radicals. These radicals exhibit high redox potential; they are more selective for oxidation by electron transfer reaction and exhibit higher activity in carbonates solution. The SO_4^- radicals could be generated by initiating SPS with a transition metal ion. The results showed that Co ions are the most efficient metal ions for the activation of $KHSO_5$, and Co_3O_4 has been used for the degradation of various phenols and dyes [29].

The degradation of SMX with time is presented in Figure 11. Two different initial concentrations of SMX, 150 and 250 μg/L, were used. At the beginning of the process, the catalyst was immersed into the SMX solution. After a period of 15 min, 1000 mg/L SPS was added into the solution, and the monitoring of degradation was started (time = 0). The concentration of SPS was chosen to be 1000 mg/L, because higher concentrations had no significant positive effect on the degradation of SMX, while lower concentrations of SPS resulted in lower degradation. Although increased oxidant concentrations will expectedly generate more radicals, these radicals in excess may suffer from partial scavenging and be converted to less reactive species [30]. In addition, higher concentrations of SPS will increase the sulfate ions on the water matrix, which is not desirable.

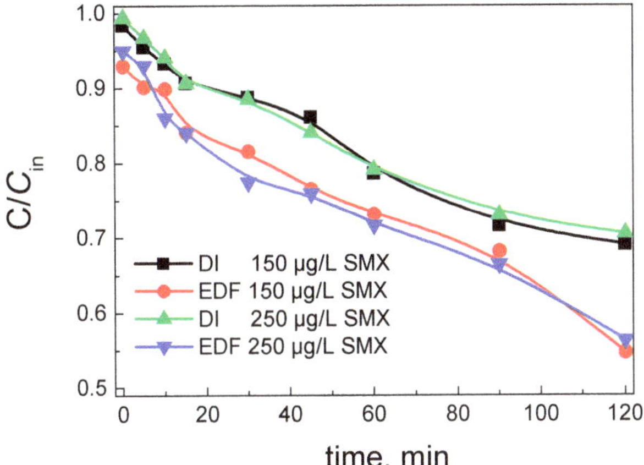

Figure 11. Degradation of sulfamethaxazole (SMX) using 1000 mg/L sodium persulfate (SPS) as oxidant and 100 mg/L catalyst with time. The concentration of SMX is presented.

As it can be seen from Figure 11, the degradation was higher for the EDF catalyst. The influence of the initial concentration was almost negligible. This is evidence that the production of radicals and the amount of adsorbed SMX were not the limiting steps under these conditions. If the rate-determining step was the production of free radicals, then the degradation would have been higher for the low concentration of SMX.

In addition, if the adsorption step was the one that determines the rate, then higher concentrations will be in favor for the degradation. In this study, the adsorption of SMX onto the catalyst surface is low, and the process is not under mass transfer phenomena. The adsorption of SMX was higher in the case of EDF, although it was still low, and almost zero for the DI catalyst. The low adsorption of SMX on the catalyst surface is rather expectable. The pzc of the catalysts were found to be 6.4 for DI and 7.6 for EDF, while the degradation process occurred at ambient pH (around 6). Generally, SMX exists as a cation only in very low pH (<1.4), while at pH >6, SMX is negatively charged, so SMX can be considered as neutral at $1.4 < pH < 5.8$ [31]. Under the pH of the degradation process, SMX and DI are almost neutral, while EDF has a slightly positive charge due to the surface sites with pH close to 5.5, making the adsorption of SMX easier; this also explains the small difference in the adsorption between the two catalysts.

The degradation data can be represented as a first-order reaction. The general kinetic model for a first order reaction rate is

$$\text{Rate} = -dC/dt = k\,C \tag{3}$$

where C is the concentration of SMX, and k is the rate constant.

The value of k can be calculated from the slope of the integrated form of Equation (3):

$$\ln C = \ln Co - k\,t. \tag{4}$$

The k value for the EDF samples was found to be equal to $(4.0 \pm 0.1) \times 10^{-3}$ min^{-1} and for the DI catalyst $(3.0 \pm 0.1) \times 10^{-3}$ min^{-1}, indicating the higher activity of the EDF catalyst, and the linear regression of the data is presented in Figure 12.

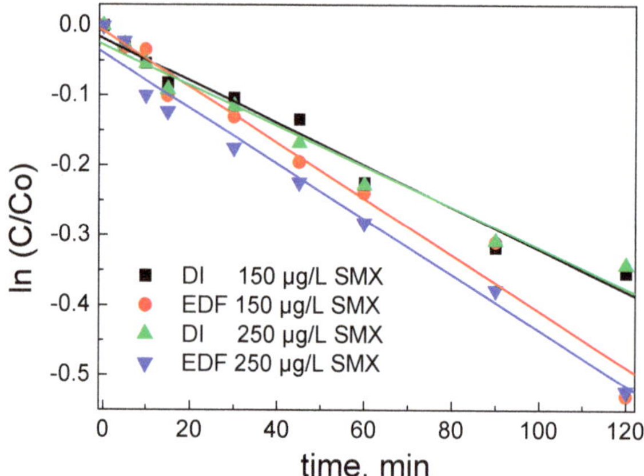

Figure 12. Linear dependence of the logarithm of SMX concentration ratio for the catalysts studied. The concentration of SMX is presented.

Higher activity of the EDF catalyst was not expected, since the EDF catalyst exhibited higher amounts of Co(III) species. According to the proposed mechanism for the activation of persulfates with Co_3O_4, the Co(II) species are the most active [32,33]. They can activate $S_2O_8^{2-}$ ions to SO_4^- radicals by the oxidation of Co(II) to Co(III), although after the reaction cycle, Co(II) regenerates. It has been shown that the surface speciation of Co_3O_4 after five reaction cycles for the Orange G degradation does not alter significantly [33]. The surface –OH are very important in the catalytic performance. According to the literature, the surface –OH groups are crucial for the catalytic performance [32]. They can help the transformation of Co(III) to Co(II). The enhanced activity of the EDF catalysts can be explained from the almost doubled surface concentration of Co species, as it was determined by XPS, and by the higher reducibility of the Co(III) species, as it was found by TPR experiments. Finally, the EDF catalyst exhibits higher interactions between the titania surface –OH groups and Co species.

3. Materials and Methods

3.1. Modification of the Typical EDF Procedure for The Preparation of Co(II)/TiO_2 Catalysts

The titania used in this work was the P25 aeroxide from Degussa (Rellinghauser Strasse 1-11, Essen 45128, Germany), and all the other chemicals used in this study were analytical grade from Merck (Merck KGaA, Headquarters of the Merck Group Frankfurter Strasse 250, Darmstadt 64293, Germany) The modified EDF procedure followed for the deposition of Co^{2+} ions on the titania involved the preparation of an aqueous solution of 0.03 M $Co(NO_3)_2$, the addition of 2 M NH_4OH solution for regulating the suspension pH of about 8.5, the impregnation by this solution of 1.0 g of TiO_2 at 25 °C, and the equilibrium of the suspension for 5 days, under nitrogen atmosphere. This helped to complete the deposition of Co(II) species and the pH to reach 7.5. The deposition of Co(II) was followed by filtration, drying at 110 °C for 2 h, and calcination in air at 400 °C for 3 h. The Co(II) concentration before and after deposition was determined in the impregnation solution using the Nitroso R-salz procedure [34]. From the difference in the Co(II) concentration before and after deposition, the Co loading was determined. The Co loading was also confirmed with desorption of the Co from the uncalcined sample and with atomic adsorption spectroscopy for a dissolved part of the final catalyst.

3.2. Physicochemical Characterization

The diffuse reflectance spectra (DRS) of the samples studied were recorded in the range 200–800 nm at room temperature, using a UV-vis spectrophotometer (Varian Cary 3, Agilent 5301 Stevens Creek

Blvd, Santa Clara, CA 95051, USA) equipped with an integration sphere. Titania was used as a reference in all the cases. The powder samples were mounted in a quartz cell, which provided a sample thickness greater than 3 mm and thus guaranteed "infinite" sample thickness. Nitrogen adsorption isotherms at liquid N_2 temperature (Tristar 3000 porosimeter Micromeritics Instrument Corp. 4356 Communications Drive, Norcross, GA 30093-2901, USA) were used for the determination of specific surface area (SSA). X-ray diffraction (XRD) patterns were recorded in a Bruker D8 (Billerica, MA, USA) Advance diffractometer equipped with a nickel-filtered CuKa (1.5418 Å) radiation source. The XPS analysis of the oxidic specimens was performed at room temperature in a UHV chamber (base pressure 8×10^{-10} mbar), which consists of fast specimen entry assembly, preparation, and an analysis chamber with residual pressure below 10^{-8} mbar, equipped with a hemispherical electron energy analyzer (SPECS, LH10) and a twin-anode X-ray gun for XPS. The unmonochromatized Mg $K\alpha$ line at 1253.6 eV and a constant pass energy mode for the analyzer were used in the experiments. The temperature-programmed reduction (TPR) experiments were performed in laboratory-constructed equipment. An amount of sample, 0.1 g, was placed in a quartz reactor, and the reducing gas mixture (H_2/Ar: 5/95 *v/v*) was passed through it for 2 h with a flow rate of 40 mL·min^{-1} at room temperature. Then, the temperature was increased to 1000 °C with a constant rate of 10 °C·min^{-1}, and the reducing gas mixture was monitoring with a thermal conductivity detector (TCD). The reducing gas mixture was dried in a cold trap (−95 °C) before reaching the TCD. To study the acid–base behavior of the Co/TiO_2-prepared catalysts, potentiometric mass titrations were applied [35]. According to PMT, the point of zero charge value of the catalyst is the common intersection point of the titration curves of suspensions with different amounts of solid and the titration curve of a blank solution. The latter is a solution that contains exactly the same amounts of inert electrolyte and base solution without solid. Applying the mass balance equation for the H^+ ions for each titration curve [36], the H^+ consumption on the biochar surface was determined. Using the differential curves of the H^+ consumption curves, valuable conclusions about the surface sites and the interfacial region can be made according to the literature [28].

3.3. *Catalytic Activity*

The prepared catalysts were tested for the degradation activity of sulfamethaxazole (SMX), which is an antibiotic using sodium persulfate (SPS) as the oxidant. Experiments were conducted in a cylindrical glass reaction vessel of 200 mL capacity, which was open to the atmosphere. An SMX solution was prepared (150 or 250 µg/L) in ultrapure water (UPW), and the appropriate amounts of SPS (1000 mg/L) and the Co/TiO_2 catalyst (100 mg/L) were added to start the reaction under magnetic stirring and ambient temperature. Samples were periodically withdrawn from the vessel and analyzed by high-performance liquid chromatography (HPLC) using an Alliance HPLC system equipped with a photodiode array detector (Waters 2996), More details about the catalytic experiments and SMX analysis can be found in [37].

4. Conclusions

Two Co/TiO_2 catalysts, with 7% CoO, were prepared using EDF and DI methods, which were characterized with various physicochemical techniques and tested in the SMX degradation. The EDF method was modified to increase the Co amount and achieved more than double Co loading. It was found that the Co loading was not the determinant factor for the activity of the catalysts. Furthermore, the production of radicals and the amount of adsorbed SMX were not the limiting steps of the reaction. The SMX degradation was found to be surface structure-sensitive, increasing with the Co surface concentration and the reducibility of the Co phase. The interactions between the Co phase and titania surface groups are important for the catalytic performance. During EDF, the Co species interact strongly with the titania surface groups, resulting in higher dispersion, a higher Co(III)/Co(II) ratio, more easily reducible Co species, and a different speciation of O surface groups, and therefore different acid–base behavior, pointing out the significance of the preparation method.

Acknowledgments: E. Siokou is gratefully acknowledged for the XPS spectra. P. Nanou is gratefully acknowledged for helping with the catalytic activity measurements.

Conflicts of Interest: The author declares no conflict of interest.

References

1. Ataloglou, T.; Vakros, J.; Bourikas, K.; Fountzoula, C.; Kordulis, C.; Lycourghiotis, A. Influence of the preparation method on the structure—Activity of cobalt oxide catalysts supported on alumina for complete benzene oxidation. *App. Catal. B Environ.* **2005**, *57*, 299–312. [CrossRef]
2. Jimenez, J.D.; Wen, C.; Lauterbach, J. Design of highly active cobalt catalysts for CO2 hydrogenation via the tailoring of surface orientation of nanostructures. *Catal. Sci. Technol.* **2019**, *9*, 1970–1978. [CrossRef]
3. Shafer, W.D.; Gnanamani, M.K.; Graham, U.M.; Yang, J.; Masuku, C.M.; Jacobs, G.; Davis, B.H. Fischer-Tropsch: Product Selectivity–The Fingerprint of Synthetic Fuels. *Catalysts* **2019**, *9*, 259. [CrossRef]
4. Petersen, A.P.; Claeys, M.; Kooyman, P.J.; van Steen, E. Cobalt-Based Fischer-Tropsch Synthesis: A Kinetic Evaluation of Metal–Support Interactions Using an Inverse Model System. *Catalysts* **2019**, *9*, 794. [CrossRef]
5. Wang, M.; Ma, J.; Yang, H.; Lu, G.; Yang, S.; Chang, Z. Nitrogen and Cobalt Co-Coped Carbon Materials Derived from Biomass Chitin as High–Performance Electrocatalyst for Aluminu—Air Batteries Catalysts. *Catalysts* **2019**, *9*, 954. [CrossRef]
6. Yung, M.M.; Holmgreen, E.M.; Ozkan, U.S. Cobalt-based catalysts supported on titania and zirconia for the oxidation of nitric oxide to nitrogen dioxide. *J. Catal.* **2009**, *247*, 356–367. [CrossRef]
7. Brik, Y.; Kacimi, M.; Ziyad, M.; Bozon-Verduraz, F. Titania-Supported Cobalt and Cobalt-Phosphorus Catalysts: Characterization and Performances in Ethane Oxidative Dehydrogenation. *J. Catal.* **2001**, *202*, 118–128. [CrossRef]
8. Yao, W.; Fang, H.; Ou, E.; Wang, J.; Yan, Z. Highly efficient catalytic oxidation of cyclohexane over cobalt-doped mesoporous titania with anatase crystalline structure. *Catal. Comm.* **2006**, *7*, 387–390. [CrossRef]
9. Eschemann, T.O.; Bitter, J.H.; de Jong, K.P. Effects of loading and synthesis method of titania–supported cobalt catalysts for Fischer–Tropsch synthesis. *Catal. Today* **2014**, *228*, 89–95. [CrossRef]
10. Bourikas, K.; Kordulis, C.; Lycourghiotis, A. The role of the liquid–solid interface in the preparation of supported catalysts. *Catal. Rev. Sci. Eng.* **2006**, *48*, 363. [CrossRef]
11. Bourikas, K.; Vakros, J.; Fountzoula, C.; Kordulis, C.; Lycourghiotis, A. Interface science for optimizing the size of oxidic nanoparticles in supported catalysts. *Catal. Today* **2007**, *128*, 138–144. [CrossRef]
12. Miller, J.T.; Schreier, M.; Kropf, A.J.; Regalbuto, J.R. A fundamental study of platinum tetraammine impregnation of silica: 2. The effect of method of preparation, loading, and calcination temperature on (reduced) particle size. *J. Catal.* **2004**, *225*, 203–212. [CrossRef]
13. Bergwerff, J.A.; Jansen, M.; Leliveld, B.G.; Visser, T.; de Jong, K.P.; Weckhuysen, B.M. Influence of the preparation method on the hydrotreating activity of MoS_2/Al_2O_3 extrudates: A Raman microspectroscopy study on the genesis of the active phase. *J. Catal.* **2006**, *243*, 292–302. [CrossRef]
14. Dumond, F.; Marceau, E.; Che, M. A Study of Cobalt Speciation in Co/Al_2O_3 Catalysts Prepared from Solutions of Cobalt−Ethylenediamine Complexes. *J. Phys. Chem. C* **2007**, *111*, 4780–4789. [CrossRef]
15. Vakros, J.; Bourikas, K.; Kordulis, C.; Lycourghiotis, A. The influence of the impregnation pH on the surface characteristics and the catalytic activity of the $Mo/\gamma\text{–}Al_2O_3$ and $CoMo/\gamma\text{–}Al_2O_3$ hydrodesulfurization catalysts prepared by Equilibrium Deposition Filtration (EDF). *J. Phys. Chem. B* **2003**, *107*, 1804–1813. [CrossRef]
16. Heracleous, E.; Vakros, J.; Lemonidou, A.A.; Kordulis, C. Role of Preparation Parameters on the Structure–Selectivity Properties of MoO_3/Al_2O_3 Catalysts for the oxidative Dehydrogenation of Ethane. *Catal. Today* **2004**, *91–92*, 289–292. [CrossRef]
17. Vakros, J.; Kordulis, C.; Lycourghiotis, A. Cobalt oxide supported γ–alumina catalyst with high active surface area prepared by equilibrium deposition filtration. *Langmuir* **2002**, *18*, 417–422. [CrossRef]
18. Petsi, T.; Panagiotou, G.D.; Garoufalis, C.S.; Kordulis, C.; Stathi, P.; Deligiannakis, Y.; Lycourghiotis, A.; Bourikas, K. Interfacial Impregnation Chemistry in the Synthesis of Cobalt Catalysts Supported on Titania. *Chem. Eur. J.* **2009**, *15*, 13090–13104. [CrossRef]

19. Vakros, J.; Bourikas, K.; Perlepes, S.; Kordulis, C.; Lycourghiotis, A. The adsorption of the cobalt ions on the "electrolytic solution/γ–alumina" interface studied by Diffuse Reflectance Spectroscopy (DRS). *Langmuir* **2004**, *20*, 10542–10550. [CrossRef]
20. Stamatis, N.; Goundani, K.; Vakros, J.; Bourikas, K.; Kordulis, C. Influence of composition and preparation method on the activity of MnO_x/Al_2O_3 catalysts for the reduction of benzaldehyde with ethanol. *Appl. Catal. A Gen.* **2007**, *325*, 322–327. [CrossRef]
21. Bourikas, K.; Fountzoula, C.; Kordulis, C. Monolayer transition metal supported on titania catalysts for the selective catalytic reduction of NO by NH_3. *Appl. Catal. B* **2004**, *52*, 145–153. [CrossRef]
22. Fountzoula, C.; Spanos, N.; Matralis, H.K.; Kordulis, C. Molybdenum-titanium oxide catalysts: The influence of the preparation conditions on their activity for the selective catalytic reduction of NO by NH_3. *Appl. Catal. B* **2002**, *35*, 295–304. [CrossRef]
23. Papavasiliou, J.; Rawski, M.; Vakros, J.; Avgouropoulos, G. A Novel Post-Synthesis Modification of CuO–CeO_2 Catalysts: Effect on Their Activity for Selective CO Oxidation. *ChemCatChem* **2018**, *10*, 2096–2106. [CrossRef]
24. Ataloglou, T.; Bourikas, K.; Vakros, J.; Kordulis, C.; Lycourghiotis, A. Kinetics of adsorption of the cobalt ions on the "electrolytic solution / γ-alumina" interface. *J. Phys. Chem. B* **2005**, *109*, 4599–4607. [CrossRef] [PubMed]
25. Voß, M.; Borgmann, D.; Wedler, G. Characterization of Alumina, Silica, and Titania Supported Cobalt Catalysts. *J. Catal.* **2002**, *212*, 10–21. [CrossRef]
26. Wei, W.F.; Chen, W.X.; Ivey, D.G. Rock Salt-Spinel Structural Transformation in Anodically Electrodeposited Mn−Co−O Nanocrystals. *Chem. Mater.* **2008**, *20*, 1941–1947. [CrossRef]
27. Zhou, W.; Wu, J.; Ouyang, C.; Gao, Y.; Xu, X.; Huang, Z. Optical properties of Mn–Co–Ni–O thin films prepared by radio frequency sputtering deposition. *J. Appl. Phys.* **2014**, *115*, 093512. [CrossRef]
28. Bourikas, K.; Kordulis, C.; Lycourghiotis, A. How metal (hydr)oxides are protonated in aqueous media: The (n + 1) rule and the role of the interfacial potential. *J. Colloid Interface Sci.* **2006**, *296*, 389–395. [CrossRef]
29. Zhang, B.-T.; Zhang, Y.; Teng, Y.; Fan, M. Sulfate Radical and Its Application in Decontamination Technologies. *Crit. Rev. Environ. Sci. Technol.* **2015**, *45*, 1756–1800. [CrossRef]
30. Fang, G.; Wu, W.; Liu, C.; Dionysiou, D.D.; Deng, Y.; Zhou, D. Activation of persulfate with vanadium species for PCBs degradation: A mechanistic study. *Appl. Catal. B Environ.* **2017**, *202*, 1–11. [CrossRef]
31. Avisar, D.; Primor, O.; Gozlan, I.; Mamane, H. Sorption of Sulfonamides and Tetracyclines to Montmorillonite Clay. *Water Air Soil Pollut.* **2010**, *209*, 439–450. [CrossRef]
32. Yang, Q.; Choi, H.; Dionysiou, D.D. Nanocrystalline cobalt oxide immobilized on titanium dioxide nanoparticles for the heterogeneous activation of peroxymonosulfate. *Appl. Catal. B Environ.* **2007**, *74*, 170–178. [CrossRef]
33. Zhang, J.; Chen, M.; Zhu, L. Activation of persulfate by Co_3O_4 nanoparticles for orange G degradation. *RSC Adv.* **2016**, *6*, 758–768. [CrossRef]
34. Samdell, E.B. *Colorimetric Determination of Traces of Metals*; Interscience: New York, NY, USA, 1950; p. 274.
35. Bourikas, K.; Vakros, J.; Kordulis, C.; Lycourghiotis, A. Potentiometric mass titrations: Experimental and theoretical establishment of a new technique for determining the point of zero charge (PZC) of metal (hydr)oxides. *J. Phys. Chem. B* **2003**, *107*, 9441–9451. [CrossRef]
36. Sfaelou, S.; Vakros, J.; Manariotis, I.D.; Karapanagioti, H.K. The use of Potentiometric Mass Titration (PMT) technique for determining the acid–base behavior of activated sludge. *Glob. NEST J.* **2015**, *17*, 397–405.
37. Magioglou, E.; Frontistis, Z.; Vakros, J.; Manariotis, I.D.; Mantzavinos, D. Activation of Persulfate by Biochars from Valorized Olive Stones for the Degradation of Sulfamethoxazole. *Catalysts* **2019**, *9*, 419. [CrossRef]

© 2020 by the author. Licensee MDPI, Basel, Switzerland. This article is an open access article distributed under the terms and conditions of the Creative Commons Attribution (CC BY) license (http://creativecommons.org/licenses/by/4.0/).

Article

Determining the Location of Co^{2+} in Zeolites by UV-Vis Diffuse Reflection Spectroscopy: A Critical View

Andrea Bellmann [1,2], Christine Rautenberg [1], Ursula Bentrup [1,*] and Angelika Brückner [1,*]

1 Leibniz-Institut für Katalyse e.V. (LIKAT) Albert-Einstein-Str. 29a, 18059 Rostock, Germany; andreabellmann89@gmail.com (A.B.); christine.rautenberg@catalysis.de (C.R.)

2 Trinseo Deutschland GmbH, Street E17, 06258 Schkopau, Germany

* Correspondence: ursula.bentrup@catalysis.de(U.B.); angelika.brueckner@catalysis.de (A.B.); Tel.: +49-381-1281-261 (U.B.)

Received: 10 December 2019; Accepted: 10 January 2020; Published: 15 January 2020

Abstract: UV–Vis spectroscopy as well as in situ FTIR spectroscopy of pyridine and CO adsorption were applied to determine the nature of Co species in microporous, mesoporous, and mixed oxide materials like Co–ZSM-5, Co/Na–ZSM-5, Co/Al–SBA-15, and Co/Al_2O_3–SiO_2. Because all sample types show comparable UV–Vis spectra with a characteristic band triplet, the former described UV–Vis band deconvolution method for determination and quantification of individual cationic sites in the zeolite appears doubtful. This is also confirmed by results of pyridine and CO adsorption revealing that all Co–zeolite samples contain two types of Co^{2+} species located at exchange positions as well as in oxide-like clusters independent of the Co content, while in Co/Al–SBA-15 and Co/Al_2O_3–SiO_2 only Co^{2+} species in oxide-like clusters occur. Consequently, the measured UV–Vis spectra represent not exclusively isolated Co^{2+} species, and the characteristic triplet band is not only related to γ-, β-, and α-type Co^{2+} sites in the zeolite but also to those dispersed on the surface of different oxide supports. The study demonstrates that for proper characterization of the formed Co species, the use of complementary methods is required.

Keywords: Co–ZSM-5; UV–Vis diffuse reflection spectroscopy; FTIR spectroscopy; pyridine adsorption; CO adsorption

---

1. Introduction

Cobalt zeolites are important catalysts for the selective catalytic reduction (SCR) of nitrogen oxides by hydrocarbons, especially by methane [1–14]. It is widely accepted that depending on the used zeolite, its Si/Al ratio, Co content, and preparation method, different Co species are formed comprising: (i) Co^{2+} cations located in the zeolite channels, (ii) CoO_x microaggregates, and (iii) CoO, Co_3O_4, as well as Co silicate mainly in the case of over-exchanged zeolites. The specific role of these species in the SCR reaction is contrarily discussed.

As Co ions are most stable in the form of Co^{2+}, two $[AlO_4]^-$ sites are needed to stabilize the charge. Thus, only Al–O–(Si–O)$_2$–Al units occurring in close proximity in the zeolite rings are able to charge balance divalent cations, while single Al atoms can only balance monovalent ions [15]. Therefore, the concentration and location of incorporated Co species are controlled by the distribution of Al atoms in the zeolite framework. Reversely, the zeolite exchange capacity for Co^{2+} can be used to probe the Al distribution in the zeolite framework [16]. Assuming a fully exchanged zeolite, the quantification of the rings containing two Al atoms in the same channel is possible. The preparation conditions to guarantee a complete exchange were described in this reference. Thus, in the group of Wichterlová [17–20], a procedure was developed enabling the discrimination between those sites using

UV–Vis diffuse reflection spectroscopy (UV–Vis–DRS). It was assumed that the d–d transition bands of bare Co^{2+} ions in the visible region of the respective dehydrated zeolites were characteristic for the individual cationic sites. Thus, the observed bands around 21,000 cm^{-1} (475 nm), 17,250 cm^{-1} (580 nm), and 16,000 cm^{-1} (625 nm) were assigned to the γ-, β-, and α-type sites, respectively. By analyzing the deconvoluted UV–Vis–DR spectra, the percentages of the respective sites occupied by Co^{2+} were determined. This approach was applied for Co–mordenite [17], Co–ferrierite [18], Co–ZSM-5 [19], and Co–beta [20]. The fundamental requirement for using this method is that all Co, represented by these UV-Vis bands, is present in form of isolated Co^{2+} and that these sites are created by wet ion exchange at a pH of 5.5 according a selected procedure [15,16].

The work of the Wichterlová group is highly recognized in the scientific community, and the developed approach to probe the Al distribution in the zeolite framework was also applied by other groups [21,22]. However, there are some doubts related to the nature of the Co sites created by ionic exchange and the assignment of the observed band triplet in the UV–Vis–DR spectra. Thus, Campa et al. [10] and Gil et al. [23,24] demonstrated that the formation of Co oxidic clusters is possible even if the formal Co/Al ratio is smaller than 0.5. This is due to the hydrolysis of the $[Co(H_2O)_6]^{2+}$ and $[Co(OH)]^+$ complexes, which proceeds easily. Hence, intermediates like $[Co–OH–Co]^{3+}$ (by olation) and $[Co–O–Co]^{2+}$ (by oxolation) can be formed which are subsequently converted into oxidic clusters during the calcination process.

On the other hand, in papers of the Busca group, [11,25,26] besides comprehensive spectroscopic characterization of Co-exchanged zeolites, UV–Vis–DR spectra of $Co/SiO_2–Al_2O_3$ samples, having an open pore structure, are also reported for comparison. These spectra were found to be very similar to those measured on Co–MFI and Co–Fer zeolites showing a triplet of band in the region 22,000–14,000 cm^{-1} too. Furthermore, as deduced from additional infrared spectroscopic studies, the presence of Co species located at the external surface of the zeolite has to be considered besides those located at exchange positions within the zeolite framework. As a consequence, both internal as well as external Co^{2+} species contribute to the UV–Vis spectrum and cannot be distinguished from each other. Therefore, it was stated that the assignment of the observed bands, being very similar for Co-exchanged MOR, FER, and MFI zeolites, to three different types of sites (α, β, γ) located in the internal cavities only is quite unreliable [26].

Inspired by these controversial statements concerning the use of UV–Vis–DRS for determining nature and distribution of Co^{2+} species in Co-exchanged zeolites and based on results of own investigations of Co–ZSM-5 catalysts used in the CH_4–SCR of NO_x, we present here a comprehensive characterization study enabling a deeper insight concerning the nature of Co species created by wet exchange procedure. Therefore, different samples were selected comprising Co–ZSM-5, Co/Na–ZSM-5, and Co/Al-SBA-15 prepared by ion exchange, as well as Co/Al_2O_3, SiO_2 prepared by incipient wetness impregnation. All samples were characterized by UV–Vis–DRS and FTIR (CO and pyridine adsorption). The aim was to demonstrate the requirement of applying complementary methods for proper characterization the formed Co species. The exchange degree reflected by the availability of acid sites was examined by pyridine adsorption, while the nature of the formed Co species (isolated and/or agglomerated) was characterized by CO adsorption at low temperature and UV–Vis–DR spectroscopy.

2. Results and Discussion

2.1. Catalysts

The catalysts together with their metal contents and the preparation methods applied in this work are listed in Table 1. Except for the 1.96 Co sample, which was prepared according to the procedure described by Dědeček et al. [19], all zeolite-based catalysts were prepared by an own method adapting the procedure from Beznis et al. [27] in which the used Co salts, the applied Co concentrations in the solutions, and temperatures for ion exchange were varied, while NH_4–ZSM-5 and Na–ZSM-5 were used as starting materials for ion exchange. Because the Si/Al ratio in the starting zeolite material was constant, the exchange degree is mainly influenced by the applied Co salt concentration. When using Na–ZSM-5 as starting material, the Na ions were not completely exchanged. For comparison, two catalysts were also prepared utilizing materials that have no microporous zeolite structure. On the one hand a mesoporous Al–SBA-15 was used for ion exchange, and on the other hand a mixed oxide SiO_2/Al_2O_3 with open pore structure was loaded with Co by incipient wetness impregnation.

Table 1. Overview of catalysts studied, applied preparation methods, and metal contents.

Catalyst	Support	Preparation Method	Content (wt%)	Co/Al
H–ZSM-5	NH_4–ZSM-5	IE: water 100 mL/1 g zeolite, 60 °C, 24 h	Si: 34.90 Al: 3.25	0.00
Na–ZSM-5	NH_4–ZSM-5	IE: 0.1 M Na—nitrate 100 mL/1 g zeolite, 25 °C, 24 h	Na: 2.37	0.00
0.32 Co	NH_4–ZSM-5	IE: 0.0005 M Co–nitrate 100 mL/1 g zeolite, 60 °C, 24 h	Co: 0.32	0.04
3.19 Co	NH_4–ZSM-5	IE: 0.05 M Co–nitrate 100 mL/1 g zeolite, 25 °C, 24 h	Co: 3.19	0.46
2.44 Co	NH_4–ZSM-5	IE: 0.005 M Co–nitrate 100 mL/1 g zeolite, 60 °C, 24 h	Co: 2.44	0.30
1.96 Co [1]	NH_4–ZSM-5	IE: 0.05 M Co–acetate 16 mL/1 g zeolite, 70 °C, 12 h	Co: 1.96	0.28
0.34 Co/1.94 Na	Na–ZSM-5	IE: 0.0005 M Co–nitrate 100 mL/1 g zeolite, 60 °C, 24 h	Co: 0.34 Na: 1.94	0.05
0.35 Co/1.98 Na	Na–ZSM-5	IE: 0.0005 M Co–acetate 100 mL/1 g zeolite, 60 °C, 24 h	Co: 0.35 Na: 1.98	0.05
2.6 Co/0.26 Na	Na–ZSM-5	IE: 0.05 M Co–nitrate 100 mL/1 g zeolite, 60 °C, 24 h	Co: 2.60 Na: 0.26	0.40
1.79 Co–SiAl	Siralox	IWI: 423 mg Co–acetate in 12.5 mL water, 5 g Siralox	Co: 1.79	0.06
0.8 Co–AlSBA	Al–SBA-15	IE: 0.02 M, Co–acetate 100 mL/1 g AlSBA, 60 °C, 24 h	Co: 0.80 Al: 2.92	0.13

[1] Preparation method according Dědeček et al. [19].

2.2. UV–Vis DRS Studies

The UV–Vis–DR spectra of the catalysts measured at room temperature after dehydration in Heat 400 °C are depicted in Figure 1. For all samples, similar spectra were obtained showing a characteristic triplet of bands with maxima around 495 nm, 585 nm, and 645 nm as earlier described in literature for Co–ZSM-5 samples [19]. The intensity ratios of these band maxima vary, as can be seen in detail by inspecting the deconvoluted spectra (Figure S1), but a clear trend in terms of Co concentration and/or used materials is not visible. Interestingly, comparable spectra with a characteristic band triplet were also obtained from the samples 1.79 Co–SiAl and 0.8 Co–AlSBA, which have no zeolite structures. This suggests, in agreement with conclusion made by the Busca group, [11,26] that the assignment of the observed band triplet to three different types of sites (α, β, γ) located in the internal cavities of pentasil-containing zeolites [17–19] might not be correct, particularly since such a characteristic band triplet is not restricted to zeolite materials and was also described, e.g., for cobalt-based blue pigments, [28] cobalt spinels, [29,30] cobalt oxide-apatite materials, [31] Co/Al_2O_3, [32], and Co-doped ZnO [33].

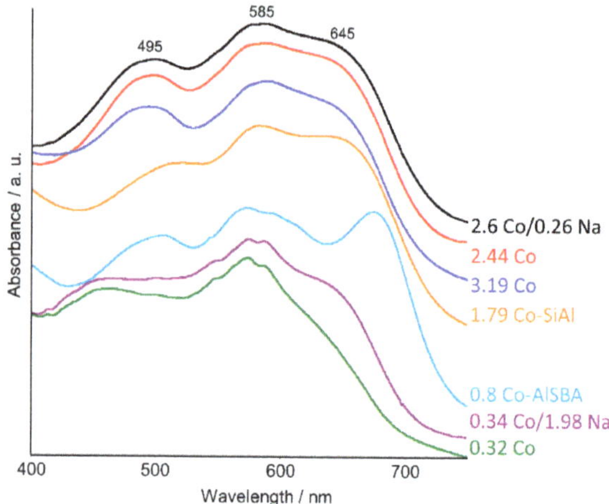

Figure 1. UV–Vis spectra measured in diffuse reflection mode at room temperature after heating the samples at 400 °C in He for 30 min.

Octahedral and tetrahedral Co^{2+} (d^7) complexes, particularly in zeolite materials, have been widely studied by optical spectroscopy in the past [34–39]. It is known that Co^{2+} changes from high-spin octahedral state in hydrated samples to tetrahedral coordination in the respective dehydrated ones, visible by the color change from pink to blue. Tetrahedral Co^{2+} (d^7) exhibits three symmetry and spin allowed transitions from the ground state: $\nu_1 = {}^4A_2(F) \rightarrow {}^4T_2(F)$, $\nu_2 = {}^4A_2(F) \rightarrow {}^4T_1(F)$, $\nu_3 = {}^4A_2(F) \rightarrow {}^4T_1(P)$ [36,39]. While ν_1 is observed in the infrared region between 2500 and 6000 cm^{-1}, the ν_2 and ν_3 transitions are observable in the near infrared and visible region. For the ν_2 and ν_3 transitions, a band splitting is observed, which may be due to (i) a low symmetry perturbation, which lifts the degeneracy of the two ${}^4T_1$ excited levels and which induces an asymmetrical band splitting; (ii) a dynamic Jahn–Teller effect, which splits the two bands into three symmetrically spaced peaks; (iii) spin-orbit coupling [38,39]. Taking these facts into account, it is not surprising that the spectra of all dehydrated Co^{2+}-containing samples, not only those of zeolites, exhibit a band triplet in the visible region between 400–750 nm.

The observed different shapes of the band triplets might by related to the specific coordination geometry of Co^{2+} influenced by the individual surrounding of the Co^{2+} ions in the respective solids. Thus, it can be expected that the splitting pattern in the case of Co-containing zeolites changes depending on the occupied positions (γ-, β-, or α-type sites), e.g., in pentasil-containing zeolites [17–20]. Therefore, one can consider that the splitting pattern provides information concerning the successive occupation of respective sites with increasing Co content as described by Dědeček et al. [19], e.g., for Co–ZSM-5 samples. Indications for that can also be seen by comparing the spectra of 0.32 Co and 2.44 Co in Figure 1. However, when Co^{2+} ions are evenly distributed on γ-, β-, and α-type sites, then each type of Co species gives rise to an own triplet band, consequently leading to three superimposed triplet bands, which contribute to the observed spectrum. Therefore, it is practically impossible to discriminate between Co species at different sites and to quantify the percentage of occupied sites by deconvolution of the observed spectra. Nevertheless, based on spectral deconvolution, Dědeček et al. [19] described different components which were ascribed to three types of Co ions located at specific cationic sites in Co–ZSM-5: A single band at 15,000 cm^{-1} for α-sites; four bands at 16,000, 17,150, 18,600, 21,200 cm^{-1} for β-sites; and a doublet at 20,100, 22,000 cm^{-1} for γ-sites. This might be mathematically correct, but not from the physical point of view. Moreover, as we have shown, the triplet bands can also be properly deconvoluted on the basis of only three components (cf. Figure S1).

It has to be mentioned here that for determining the location of the Co^{2+} ions, it was presupposed [15,19] that all Co^{2+} ions are isolated due to the exchange of protons (Brønsted sites), which are created by the incorporation of Al in the zeolite matrix. Hence, only the Co/Al ratio was taken into account. We demonstrate in the next paragraph that this supposition cannot be made without inclusion of additional characterization methods.

2.3. FTIR Studies: Pyridine and CO Adsorption

Brønsted and Lewis acidic sites can be well characterized by FTIR spectroscopic investigation of pyridine adsorption [40,41]. The typical bands resulting from pyridine adsorbed on Brønsted acid sites (PyH^+) are observed around 1540 cm^{-1}, while the bands from pyridine interacting with Lewis acid sites (L–Py) are detected at around 1450 cm^{-1} and in the region of 1625–1595 cm^{-1}. From the band positions in the latter region, the strength of Lewis acid sites can be evaluated [41]. The pyridine adsorbate spectra of the studied catalysts measured at 150 °C are depicted in Figure 2. As expected, the Co-free H–ZSM-5 sample shows an intense PyH^+ band at 1545 cm^{-1}, the intensity of which decreases with Co loading. Upon comparing the band intensities of the PyH^+ band, it is obvious that even the highly loaded sample 3.19 Co possesses a quite high number of Brønsted acid sites, although the Co/Al ratio is very close to 0.5. This means that a part of the Co^{2+} ions is obviously not located on exchange positions, which points to an exchange process with a more complex stoichiometry as also observed by other authors [10].

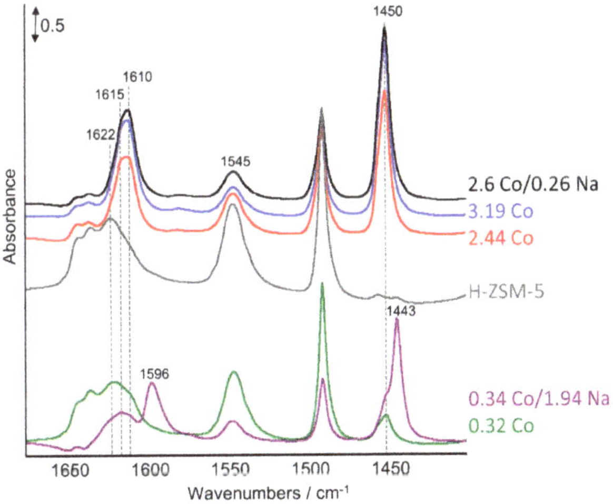

Figure 2. Pyridine adsorbate spectra measured at 150 °C.

By comparing the intensities of the L–Py band of H–ZSM-5 with that of the Co-containing samples, it is seen that additional Lewis acid sites are created with increasing Co content as indicated by the bands at 1450/1610 cm^{-1} (Co^{2+}) and 1443/1596 cm^{-1} (Na^+). Furthermore, a splitting of the L-Py band in the region above 1600 cm^{-1} into two components at 1610 and 1616 cm^{-1} is observable, revealing two Co^{2+} Lewis sites of different strength, possibly located at different sites in the zeolite lattice.

A suitable method for characterizing the nature of Co species is adsorption of CO at low temperatures [12,13,42–44]. Thus, one can distinguish Co^{2+} located at exchange positions from Co^{2+} being part of oxide-like clusters because the respective νCo^{2+}–CO bands appear at different positions, namely at 2204 cm^{-1} for Co^{2+} in exchange positions and at 2194 cm^{-1} for Co^{2+} in oxide-like clusters. The CO adsorbate spectra of the studied samples are compared in Figure 3. Here, the spectra measured at −60 °C are shown because the band stemming from the interaction of CO with OH groups at 2174 cm^{-1} [45] has negligible intensity at this temperature. This band is intense at the adsorption temperature of −120 °C but vanishes at higher temperatures because the interaction of CO with hydroxyl groups is not as strong as the interaction with Co^{2+} ions [14]. It is obvious that all Co–zeolite samples (Figure 3a) contain Co^{2+} species at exchange positions (2204 cm^{-1}) as well as in oxide-like clusters (2195 cm^{-1}) independent from the Co content. This is in accordance with the results of pyridine adsorption, where two Co^{2+} Lewis sites of different strength have been identified. The bands at 2177 and 2112 cm^{-1} in the 0.35 Co/1.98 Na sample are related to νNa^{+}–CO and νNa^{+}–OC vibrations, respectively [46].

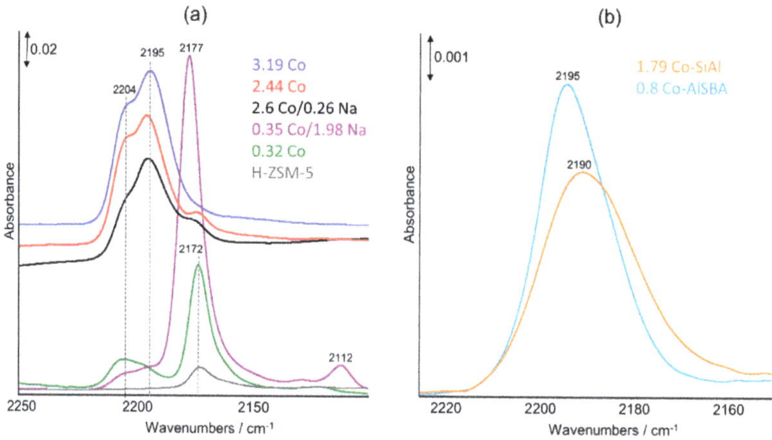

Figure 3. CO adsorbate spectra of Co-ZSM-5 samples (**a**) and 1.79 Co-SiAl as well as 0.8 Co-AlSBA (**b**) measured at −60 °C.

In the CO adsorbate spectra of 1.79 Co–SiAl and 0.8 Co–AlSBA (Figure 3b), only Co^{2+} in oxide-like clusters is observed as reflected by the band position. With respect to the UV–Vis spectra, which are similar for both types of samples (cf. Figure 1), the results of CO adsorption lead to the conclusion that the triplet bands observed in the spectra are related to tetrahedral Co^{2+} sitting at exchange positions as well as in oxide-like clusters, which can be located inside as well as outside the zeolite channels, depending on the cluster size. The latter conclusion is based on CO adsorption experiments in which o-toluonitrile was preadsorbed on Co–H–MFI [12,13]. In this way, the discrimination between cationic sites inside and outside the channels was possible because the bulky molecule is not able to penetrate the channels of zeolites like ZSM-5 or MOR and therefore block respective sites. Thus, it could be demonstrated for Co–H–MFI that Co^{2+} ions are distributed at both the external and internal surfaces. This has to be assumed too for the Co–ZSM-5 catalysts studied in this work.

To be sure that the findings of pyridine and CO adsorption are not influenced by the used preparation methods, being different from that described by Dědeček et al. [19] we have also prepared a sample according to this protocol. As a result, the sample 1.96 Co was obtained (cf. Table 1), the characterization results of which in comparison with those of 2.44 Co are shown in Figure 4.

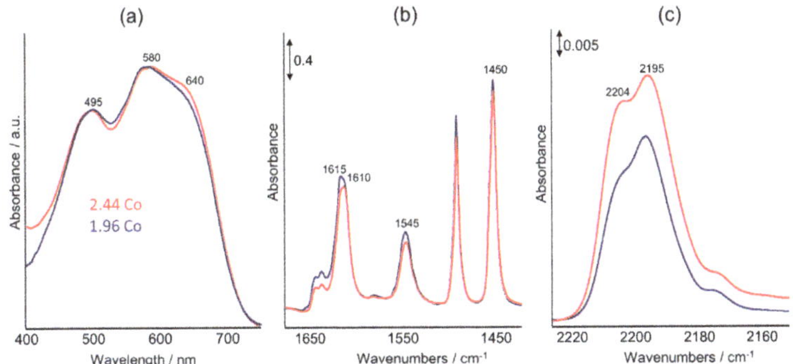

Figure 4. Comparison of UV–Vis–DR spectra measured at room temperature after heating the samples at 400 °C in He for 30 min (**a**), pyridine adsorbate spectra measured at 150 °C (**b**), and CO adsorbate spectra measured at −50 °C (**c**) of the samples 1.96 Co and 2.44 Co.

It is seen that the UV–Vis–DR as well as the pyridine and CO adsorbate spectra are very similar. While the UV–Vis spectra are nearly identical (Figure 4a), the Brønsted acidity of the 1.96 Co sample is slightly higher compared to that of 2.44 Co (Figure 4b), which might be due to the lower Co content and therefore lower exchange level. This is also true for the intensities of the observed CO adsorbate bands, whereas the intensity ratio of the bands at 2204/2195 cm^{-1} is nearly the same. In summary, it can be stated that in both catalysts, independent of the applied exchange procedure, Co^{2+} species are located both at exchange positions and in oxide-like clusters. Hence, as already mentioned above, the respective UV–Vis spectra reflect not only Co^{2+} ions located at α, β, γ sites in the internal cavities of the zeolite but also those in oxide-like clusters inside and/or outside the zeolite channels.

The existence of oxide-like clusters implies the possible presence of Co^{3+} besides Co^{2+} ions. Considering the fact that the samples were calcined in synthetic air at 500 °C, the presence of Co^{3+} has to be taken into account, which was also proven by NO adsorption in a former study [14]. For investigating the possible influence of calcination conditions on the nature of formed Co species, we analyzed sample 1.96 Co sample by respective FTIR experiments with CO and exemplarily the 2.44 Co catalyst by UV–Vis–DRS.

In Figure 5a, the normalized CO adsorbate spectra of 1.96 Co are shown measured at −60 °C after pretreatment of the fresh, non-calcined sample in situ in the FTIR reaction cell by heating in vacuum for 1 h at 500 °C (vac.) and heating in synthetic air at 500 °C for 45 min (air). For comparison, the spectrum of the calcined (calc.) sample is also shown. The spectrum of the latter is similar to that of the vacuum-treated sample exhibiting the typical bands of CO adsorbed on Co^{2+} at exchange positions as well as Co^{2+} in oxide-like clusters (vide supra). In the spectrum of the air-treated sample, only Co^{2+} in oxide-like clusters is detectable. The high intensity of the νOH–CO band around 2172 cm^{-1} in this sample in relation to the respective ν Co^{2+}–CO bands suggests an agglomeration of the Co^{2+} species that recreates Brønsted sites, i.e., not exchanged protons.

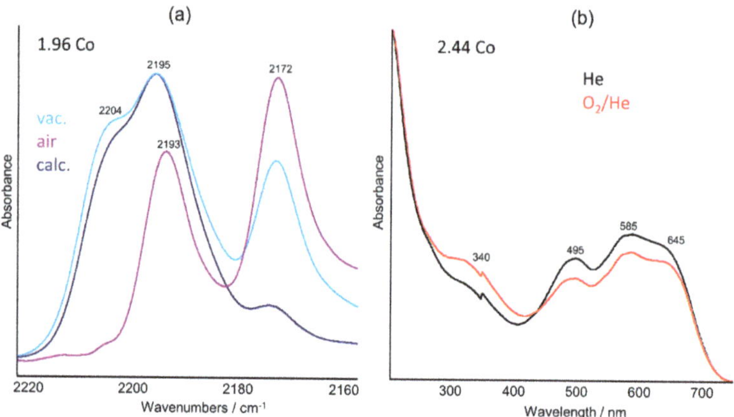

Figure 5. CO adsorbate normalized spectra measured at −60 °C (**a**) of the samples 1.96 Co and (**b**) 2.44 Co.

Figure 5b shows the UV–Vis spectra of the sample 2.44 Co measured at room temperature after heating at 400 °C for 60 min followed by heating in 2 vol% O_2/He at 400 °C for 45 min. After oxidation a new band around 340 nm appears, which is attributed to a ligand to metal charge transfer transition from framework oxide ions to Co^{3+} ions [37,39,47]. In parallel, the intensity of the triplet band of Co^{2+} decreases accordingly. This means that a part of Co^{2+} species, most probably those occurring in oxide-like clusters, has been oxidized to Co^{3+}.

These experiments reveal that the calcination/dehydration conditions influence the nature and location of formed Co species, but even treatment in vacuum at high temperature does not inevitably lead to Co^{2+} located only at exchange positions. The fact that part of the Co^{2+} species can be easily oxidized leads to the conclusion that these Co^{2+} ions occur mainly in oxide-like agglomerates located inside and/or outside the zeolite channels. This means in terms of the UV–Vis spectra of such samples that the observed spectra represent not exclusively isolated Co^{2+} species but also those located in agglomerated, oxide-like species inside and/or outside the zeolite channels. The same conclusion was drawn from the results of pyridine and CO adsorption as described above.

These findings are in accordance with those from former studies [12,25,26] revealing that the distribution of Co species is by far more complex, as deduced from previous papers in which UV–Vis, EXAFS, and XRD data have been discussed [17–19,48]. Insofar, a detailed analysis and quantification of Co positions in the zeolite matrix on the basis of deconvoluted UV–Vis spectra is practically not possible because the spectrum reflects Co species occupying positions inside the zeolite channels as well as positions on internal and external surfaces. This conclusion is also supported by the fact that for Co sites in zeolite cavities as well as for those dispersed on the surface of different oxide supports, similar UV–Vis spectra with the characteristic band triplet are obtained. The characteristic band splitting is due to symmetry perturbation, dynamic Jahn–Teller effect, and spin-orbit coupling [38,39] and is observed in general for Co^{2+} ions independently from the surrounding matrix. For determining nature and distribution of Co^{2+} species in Co-exchanged zeolites, the use of only UV–Vis–DRS cannot reflect the real situation. For this purpose, the combination of several, complementary methods have to be applied.

From the presented results the following conclusions can be derived. The observed triplet bands in the UV–Vis spectra of Co^{2+}-containing samples reflect the sum of spectra of all existing Co^{2+} species (isolated and oxide-like) located at different positions inside the zeolite channels as well as on internal and external surfaces. The characteristic band splitting is due to symmetry perturbation, dynamic Jahn–Teller effect, and spin-orbit coupling and is generally observed for Co^{2+} ions in zeolite as well as oxide materials. Because comparable UV–Vis spectra with a characteristic band triplet is observed for Co^{2+} ions independently from the surrounding matrix, the former described UV–Vis band deconvolution method for determination and quantification of individual cationic sites in zeolites is not applicable.

3. Materials and Methods

The Co–ZSM-5 catalysts were prepared starting from NH_4–ZSM-5 (Zeolyst, CBV 2314, SiO_2/Al_2O_3 = 23). Co was introduced by wet ion-exchange (IE) using solutions of $Co(NO_3)_2$–$6H_2O$ (Sigma-Aldrich, St. Louis, MO, USA) or $Co(CH_3COO)_2$–$4H_2O$ (Sigma-Aldrich, St. Louis, MO, USA), adapting the method described by Beznis et al. [27]. The used conditions are summarized in Table 1. The zeolite was suspended in the respective Co salt aqueous solution and preheated to 60 °C. The pH was adjusted to 6.9, and the solution was stirred for 24 h. After filtration, the material was washed with 400 mL desalinated water/g solid. After drying over night at 90 °C, the solids were calcined in synthetic air at 500 °C for 6 h. For the preparation of the Na–ZSM-5, a solution of $NaNO_3$ (Sigma-Aldrich, St. Louis, MO, USA) was used. The exchange procedure and subsequent washing was repeated two times. Finally, the solid was dried at 100 °C for 24 h and calcined in synthetic air at 500 °C for 6 h. For comparison, the respective H–ZSM-5 catalyst was prepared applying the same IE method but using only water. The sample 1.96 Co was prepared by the method described by Dědeček et al. [19] with adjusted pH of 5.5.

The Co–SiAl sample was prepared by incipient wetness impregnation (IWI) of a Siralox sample from Sasol (49.2% SiO_2, 50.8% Al_2O_3), which was dried overnight at 120 °C. After dropwise adding of a respective volume of a solution of $Co(CH_3COO)_2{\cdot}4H_2O$ (Carl Roth, Karlsruhe, Germany) under stirring and further stirring for 2 h, the suspension was dried overnight and calcined at 500 °C for 6 h in synthetic air.

For the preparation of Co–AlSBA, the starting Al–SBA-15 material was prepared according the method described by Wu et al. [49]. The obtained solid was dried for 12 h at 70 °C and then calcined in synthetic air at 650 °C for 16 h. This Al–SBA-15 (Si/Al = 20) was used for the IE procedure described for Co–ZSM-5 above.

The Si/Al ratio of H–ZSM-5 and the contents of Co, Al, Na were determined by ICP-OES analysis.

The UV–Vis–DR spectra were recorded in diffuse reflection mode between 200 and 800 nm on a Cary 5000 spectrophotometer (Agilent, Santa Clara, CA, USA) equipped with a diffuse reflection accessory (Praying mantis) and a Harrick reaction cell. After treatment, the samples were heated at 400 °C for 30 min, and the spectra were measured at room temperature.

In situ FTIR measurements were carried out in transmission mode on a Bruker Tensor 27 spectrometer or Nicolet iS10 spectrometer (Thermo Fisher Scientific, Waltham, MA, USA) equipped with custom-made reaction cells with CaF_2 windows for adsorption studies of pyridine and CO, respectively. The sample powders were pressed into self-supporting wafers with a diameter of 20 mm and a weight of 50 mg. For each experiment, the sample was pretreated in synthetic air at 400 °C for 1 h. After cooling down to the respective adsorption temperature and evacuation of the cell a background, spectrum of the sample was recorded. Pyridine was adsorbed at 150 °C temperature until saturation. Then, the reaction cell was evacuated to remove physisorbed pyridine, and an adsorbate spectrum was recorded. For CO adsorption, a 5% CO/He mixture was used, dosed in pulses at −120 °C until saturation. To remove the physisorbed CO, the cell was evacuated, and the CO adsorbate spectrum was recorded. Subsequently, the CO desorption under vacuum was followed by continuously heating the sample and measuring a spectrum every 10 °C. Generally, the spectrum of the catalyst measured after pretreatment at adsorption temperature was subtracted from the respective adsorbate spectra of pyridine and CO.

Supplementary Materials: The following are available online at http://www.mdpi.com/2073-4344/10/1/123/s1, Figure S1: Deconvoluted UV-Vis–DR spectra of selected samples measured at room temperature after heating the samples at 400 °C in He for 30 min.

Author Contributions: Conceptualization, U.B. and A.B.; investigation, A.B. and C.R.; data curation, A.B. and C.R.; writing—original draft preparation, U.B.; writing—review and editing, A.B.; supervision, A.B.; funding acquisition, U.B. All authors have read and agreed to the published version of the manuscript.

Funding: This research was funded by the Leibniz-Gemeinschaft, grant number SAW-2013-LIKAT-2.

Acknowledgments: The authors thank Anja Simmula for the ICP-OES analysis and Dominik Seeburg for the preparation of Al–SBA-15.

Conflicts of Interest: The authors declare no conflict of interest.

References

1. Armor, J.N. Catalytic reduction of nitrogen-oxides with methane in the presence of excess oxygen–A review. *Catal. Today* **1995**, *26*, 147–158. [CrossRef]
2. Lonyi, F.; Solt, H.E.; Paszti, Z.; Valyon, J. Mechanism of NO-SCR by methane over Co,H–ZSM-5 and Co,H-mordenite catalysts. *Appl. Catal. B Environ.* **2014**, *150*, 218–229. [CrossRef]
3. Wang, X.; Chen, H.Y.; Sachtler, W.M.H. Catalytic reduction of NOx by hydrocarbons over Co/ZSM-5 catalysts prepared with different methods. *Appl. Catal. B Environ.* **2000**, *26*, L227–L239. [CrossRef]
4. Wang, X.; Chen, H.-Y.; Sachtler, W.M.H. Mechanism of the selective reduction of NOx over Co/MFI: Comparison with Fe/MFI. *J. Catal.* **2001**, *197*, 281–291. [CrossRef]
5. Chupin, C.; Vanveen, A.; Konduru, M.; Despres, J.; Mirodatos, C. Identity and location of active species for NO reduction by CH4 over Co-ZSM-5. *J. Catal.* **2006**, *241*, 103–114. [CrossRef]
6. Sadovskaya, E.M.; Suknev, A.P.; Pinaeva, L.G.; Goncharov, V.B.; Bal'zhinimaev, B.S.; Chupin, C.; Pérez-Ramírez, J.; Mirodatos, C. Mechanism and kinetics of the selective NO reduction over Co-ZSM-5 studied by the SSITKA technique: 2. Reactivity of NOx-adsorbed species with methane. *J. Catal.* **2004**, *225*, 179–189. [CrossRef]
7. Sun, T.; Fokema, M.D.; Ying, J.Y. Mechanistic study of NO reduction with methane over Co^{2+} modified ZSM-5 catalysts. *Catal. Today* **1997**, *33*, 251–261. [CrossRef]
8. Campa, M.C.; DeRossi, S.; Ferraris, G.; Indovina, V. Catalytic activity of Co-ZSM-5 for the abatement of NOx with methane in the presence of oxygen. *Appl. Catal. B Environ.* **1996**, *8*, 315–331. [CrossRef]
9. Campa, M.C.; Indovina, V. Cobalt-exchanged mordenites: Preparation, characterization and catalytic activity for the abatement of NO with CH_4 in the presence of excess O-2. *J. Porous Mater.* **2007**, *14*, 251–261. [CrossRef]
10. Campa, M.C.; Luisetto, I.; Pietrogiacomi, D.; Indovina, V. The catalytic activity of cobalt-exchanged mordenites for the abatement of NO with CH_4 in the presence of excess O-2. *Appl. Catal. B Environ.* **2003**, *46*, 511–522. [CrossRef]
11. Resini, C.; Montanari, T.; Nappi, L.; Bagnasco, G.; Turco, M.; Busca, G.; Bregani, F.; Notaro, M.; Rocchini, G. Selective catalytic reduction of NOx by methane over Co-H-MFI and Co-H-FER zeolite catalysts: Characterisation and catalytic activity. *J. Catal.* **2003**, *214*, 179–190. [CrossRef]
12. Montanari, T.; Marie, O.; Daturi, M.; Busca, G. Searching for the active sites of Co-H-MFI catalyst for the selective catalytic reduction of NO by methane: A FT-IR in situ and operando study. *Appl. Catal. B Environ.* **2007**, *71*, 216–222. [CrossRef]
13. Montanari, T.; Marie, O.; Daturi, M.; Busca, G. Cobalt on and in zeolites and silica–alumina: Spectroscopic characterization and reactivity. *Catal. Today* **2005**, *110*, 339–344. [CrossRef]
14. Bellmann, A.; Atia, H.; Bentrup, U.; Brückner, A. Mechanism of the selective reduction of NOx by methane over Co-ZSM-5. *Appl. Catal. B Environ.* **2018**, *230*, 184–193. [CrossRef]
15. Dedecek, J.; Sobalik, Z.; Wichterlova, B. Siting and distribution of framework aluminium atoms in silicon-rich zeolites and impact on catalysis. *Catal. Rev.* **2012**, *54*, 135–223. [CrossRef]
16. Dědeček, J.; Kaucký, D.; Wichterlová, B.; Gonsiorová, O. Co^{2+} ions as probes of Al distribution in the framework of zeolites. ZSM-5 study. *Phys. Chem. Chem. Phys.* **2002**, *4*, 5406–5413.
17. Dědeček, J.; Wichterlová, B. Co^{2+}Ion siting in pentasil-containing zeolites. I. Co^{2+}Ion sites and their occupation in mordenite. A Vis–NIR diffuse reflectance spectroscopy study. *J. Phys. Chem. B* **1999**, *103*, 1462–1476.
18. Kaucký, D.; Dědeček, J.; Wichterlová, B. Co^{2+} ion siting in pentasil-containing zeolites. *Microporous Mesoporous Mater.* **1999**, *31*, 75–87. [CrossRef]

19. Dedecek, J.; Kaucky, D.; Wichterlova, B. Co^{2+} ion siting in pentasil-containing zeolites, part 3. Co^{2+} ion sites and their occupation in ZSM-5: A VIS diffuse reflectance spectroscopy study. *Microporous Mesoporous Mater.* **2000**, *35–36*, 483–494. [CrossRef]
20. Dědeček, J.; Čapek, L.; Kaucký, D.; Sobalík, Z.; Wichterlová, B. Siting and distribution of the Co Ions in beta zeolite: A UV–Vis–NIR and FTIR study. *J. Catal.* **2002**, *211*, 198–207. [CrossRef]
21. Kim, J.; Jentys, A.; Maier, S.M.; Lercher, J.A. Characterization of Fe-exchanged BEA zeolite under NH_3 selective catalytic reduction conditions. *J. Phys. Chem. C* **2013**, *117*, 986–993. [CrossRef]
22. Kim, S.; Park, G.; Woo, M.H.; Kwak, G.; Kim, S.K. Control of hierarchical structure and framework-Al distribution of ZSM-5 via adjusting crystallization temperature and their effects on methanol conversion. *ACS Catal.* **2019**, *9*, 2880–2892. [CrossRef]
23. Gil, B.; Janas, J.; Włoch, E.; Olejniczak, Z.; Datka, J.; Sulikowski, B. The influence of the initial acidity of HFER on the status of Co species and catalytic performance of CoFER and InCoFER in CH4-SCR-NO. *Catal. Today* **2008**, *137*, 174–178. [CrossRef]
24. Gil, B.; Pietrzyk, P.; Datka, J.; Kozyra, P.; Sojka, Z. Speciation of cobalt in CoZSM-5 upon thermal treatment. In *Studies in Surface Science and Catalysis*; Čejka, J., Žilková, N., Nachtigall, P., Eds.; Elsevier: Amsterdam, The Netherlands, 2005; Volume 158, pp. 893–900.
25. Bagnasco, G.; Turco, M.; Resini, C.; Montanari, T.; Bevilacqua, M.; Busca, G. On the role of external Co sites in NO oxidation and reduction by methane over Co?H-MFI catalysts. *J. Catal.* **2004**, *225*, 536–540. [CrossRef]
26. Montanari, T.; Bevilacqua, M.; Resini, C.; Busca, G. UV–Vis and FT-IR study of the nature and location of the active sites of partially exchanged Co–H zeolites. *J. Phys. Chem. B* **2004**, *108*, 2120–2127. [CrossRef]
27. Beznis, N.V.; Weckhuysen, B.M.; Bitter, J.H. Partial oxidation of methane over Co-ZSM-5: Tuning the oxygenate selectivity by altering the preparation route. *Catal. Lett.* **2010**, *136*, 52–56. [CrossRef]
28. Llusar, M.; Forés, A.; Badenes, J.A.; Calbo, J.; Tena, M.A.; Monrós, G. Colour analysis of some cobalt-based blue pigments. *J. Eur. Ceram. Soc.* **2001**, *21*, 1121–1130. [CrossRef]
29. Lee, K.; Ruddy, D.A.; Dukovic, G.; Neale, N.R. Synthesis, optical, and photocatalytic properties of cobalt mixed-metal spinel oxides $Co(Al_{1-x}Ga_x)_2O_4$. *J. Mater. Chem. A* **2015**, *3*, 8115–8122. [CrossRef]
30. Rangappa, D.; Ohara, S.; Naka, T.; Kondo, A.; Ishii, M.; Adschiri, T. Synthesis and organic modification of $CoAl_2O_4$ nanocrystals under supercritical water conditions. *J. Mater. Chem.* **2007**, *17*, 4426–4429. [CrossRef]
31. El Kabouss, K.; Kacimi, M.; Ziyad, M.; Ammar, S.; Ensuque, A.; Piquemal, J.-Y.; Bozon-Verduraz, F. Cobalt speciation in cobalt oxide-apatite materials: Structure–properties relationship in catalytic oxidative dehydrogenation of ethane and butan-2-ol conversion. *J. Mater. Chem.* **2006**, *16*, 2453–2463. [CrossRef]
32. Liotta, L.F.; Pantaleo, G.; Macaluso, A.; Di Carlo, G.; Deganello, G. CoOx catalysts supported on alumina and alumina-baria: Influence of the support on the cobalt species and their activity in NO reduction by C3H6 in lean conditions. *Appl. Catal. A Gen.* **2003**, *245*, 167–177. [CrossRef]
33. Radovanovic, P.V.; Norberg, N.S.; McNally, K.E.; Gamelin, D.R. Colloidal transition-metal-doped ZnO quantum dots. *J. Am. Chem. Soc.* **2002**, *124*, 15192–15193. [CrossRef] [PubMed]
34. Cotton, F.A.; Goodgame, D.M.L.; Goodgame, M. The electronic structures of tetrahedral cobalt(II) complexes *J. Am. Chem. Soc.* **1961**, *83*, 4690–4699. [CrossRef]
35. Klier, K.; Kellerman, R.; Hutta, P.J. Spectra of synthetic zeolites containing transition metal ions. V.* Π complexes of olefins and acetylene with Co(II)A molecular sieve. *J. Chem. Phys.* **1974**, *61*, 4224–4234. [CrossRef]
36. Klier, K. Transition-metal ions in zeolites: The perfect surface sites. *Langmuir* **1988**, *4*, 13–25. [CrossRef]
37. Verberckmoes, A.A.; Uytterhoeven, M.G.; Schoonheydt, R.A. Framework and extra-framework Co2+ in CoAPO-5 by diffuse reflectance spectroscopy. *Zeolites* **1997**, *19*, 180–189. [CrossRef]
38. Praliaud, H.; Coudurier, G. Optical spectroscopy of hydrated, dehydrated and ammoniated cobalt(II) exchanged zeolites X and Y. *J. Chem. Soc. Faraday Trans. 1 Phys. Chem. Condens. Phases* **1979**, *75*, 2601–2616. [CrossRef]
39. Verberckmoes, A.A.; Weckhuysen, B.M.; Schoonheydt, R.A. Spectroscopy and coordination chemistry of cobalt in molecular sieves. *Microporous Mesoporous Mater.* **1998**, *22*, 165–178. [CrossRef]
40. Buzzoni, R.; Bordiga, S.; Ricchiardi, G.; Lamberti, C.; Zecchina, A.; Bellussi, G. Interaction of pyridine with acidic (H–ZSM5, H-β, H-MORD Zeolites) and superacidic (H-Nafion Membrane) systems: An IR investigation. *Langmuir* **1996**, *12*, 930–940. [CrossRef]

41. Busca, G. The surface acidity of solid oxides and its characterization by IR spectroscopic methods. An attempt at systematization. *Phys. Chem. Chem. Phys.* **1999**, *1*, 723–736. [CrossRef]
42. Gora-Marek, K.; Gil, B.; Sliwa, M.; Datka, J. An IR spectroscopy study of Co sites in zeolites CoZSM-5. *Appl. Catal. A Gen.* **2007**, *330*, 33–42. [CrossRef]
43. Góra-Marek, K.; Gil, B.; Datka, J. Quantitative IR studies of the concentration of Co^{2+} and Co^{3+} sites in zeolites CoZSM-5 and CoFER. *Appl. Catal. A Gen.* **2009**, *353*, 117–122. [CrossRef]
44. Gora-Marek, K. The reduction and oxidation of Co species in CoZSM-5 zeolites studied by IR spectroscopy. *Top. Catal.* **2009**, *52*, 1023–1029. [CrossRef]
45. Kustov, L.M.; Kazanskii, V.B.; Beran, S.; Kubelkova, L.; Jiru, P. Adsorption of carbon monoxide on ZSM-5 zeolites: Infrared spectroscopic study and quantum-chemical calculations. *J. Phys. Chem.* **1987**, *91*, 5247–5251. [CrossRef]
46. Zecchina, A.; Bordiga, S.; Lamberti, C.; Spoto, G.; Carnelli, L.; Otero Arean, C. Low-temperature fourier transform infrared study of the interaction of CO with cations in alkali-metal exchanged ZSM-5 zeolites. *J. Phys. Chem.* **1994**, *98*, 9577–9582. [CrossRef]
47. Kraushaar-Czarnetzki, B.; Hoogervorst, W.G.M.; Andréa, R.R.; Emeis, C.A.; Stork, W.H.J. Characterisation of CoIIand CoIIIin CoAPO molecular sieves. *J. Chem. Soc. Faraday Trans.* **1991**, *87*, 891–895. [CrossRef]
48. Drozdová, L.; Prins, R.; Dědeček, J.; Sobalík, Z.; Wichterlová, B. Bonding of Co ions in ZSM-5, ferrierite, and mordenite: An X-ray Absorption, UV–Vis, and IR study. *J. Phys. Chem. B* **2002**, *106*, 2240–2248. [CrossRef]
49. Wu, S.; Han, Y.; Zou, Y.-C.; Song, J.-W.; Zhao, L.; Di, Y.; Liu, S.-Z.; Xiao, F.-S. Synthesis of heteroatom substituted SBA-15 by the "pH-Adjusting" method. *Chem. Mater.* **2004**, *16*, 486–492. [CrossRef]

© 2020 by the authors. Licensee MDPI, Basel, Switzerland. This article is an open access article distributed under the terms and conditions of the Creative Commons Attribution (CC BY) license (http://creativecommons.org/licenses/by/4.0/).

MDPI
St. Alban-Anlage 66
4052 Basel
Switzerland
Tel. +41 61 683 77 34
Fax +41 61 302 89 18
www.mdpi.com

Catalysts Editorial Office
E-mail: catalysts@mdpi.com
www.mdpi.com/journal/catalysts

www.ingramcontent.com/pod-product-compliance
Lightning Source LLC
LaVergne TN
LVHW070627170726
843515LV00004B/197

* 9 7 8 3 0 3 6 5 1 4 2 2 2 *